Joshua M. Pearce (Ed.)

Photovoltaic Materials and Electronic Devices

MDPI

This book is a reprint of the Special Issue that appeared in the online, open access journal, *Materials* (ISSN 1996-1944) from 2015–2016 (available at: http://www.mdpi.com/journal/materials/special_issues/Photovoltaic-Materials-and-Electronic-Devices).

Guest Editor
Joshua M. Pearce
Michigan Technological University
USA

Editorial Office
MDPI AG
Klybeckstrasse 64
Basel, Switzerland

Publisher
Shu-Kun Lin

Managing Editor
Leo Jiang

1. Edition 2016

MDPI • Basel • Beijing • Wuhan • Barcelona

ISBN 978-3-03842-216-7 (Hbk)
ISBN 978-3-03842-217-4 (PDF)

Table of Contents

List of Contributors

Bilel Azeza Laboratoire Micro-Optoélectroniques et Nanostructures, Faculté des Sciences de Monastir, Université de Monastir, Monatir 5019, Tunisia; Turaif Sciences College, Northern Borders University, P.O. 833, Turaif 91411, Kingdom of Saudi Arabia.

Nadia Barbero Department of Chemistry and NIS Interdepartmental Centre, University of Torino, Via Giuria 7, I-10125 Torino, Italy.

Marco Barink TNO Applied Sciences, High Tech Campus 21, Eindhoven 5656 AE, The Netherlands.

Claudia Barolo Department of Chemistry and NIS Interdepartmental Centre, University of Torino, Via Giuria 7, I-10125 Torino, Italy.

Patrick K. Bowen Department of Materials Science & Engineering, Michigan Technological University, 1400 Townsend, Houghton, MI 49931, USA.

Michael Clavel Bradley Department of Electrical and Computer Engineering, Virginia Polytechnic Institute State University, 302 Whittemore Hall, VA 24061, USA.

Robert W. Collins Wright Center for Photovoltaics Innovation & Commercialization and Department of Physics & Astronomy, University of Toledo, Toledo, OH 43606, USA.

Walid A. Daoud School of Energy and Environment, City University of Hong Kong, Tat Chee Avenue, Hong Kong, China.

Afif Fouzri Laboratoire de Physico-Chimie des Matériaux, Faculté des Sciences de Monastir, Université de Monastir, Monastir 5019, Tunisia.

Soha Gaballah Center of Smart Nanotechnology and Photonics (CSNP), Smart Critical Infrastructure (SmartCI) Research Center, Alexandria University, Elhadara, Alexandria 21544, Egypt; Department of Chemical Engineering, Faculty of Engineering, Alexandria University, Elhadara, Alexandria 21544, Egypt.

Jephias Gwamuri Department of Materials Science & Engineering, Michigan Technological University, 1400 Townsend, Houghton, MI 49931, USA.

Mohamed Helmi Hadj Alouane Laboratoire Micro-Optoélectroniques et Nanostructures, Faculté des Sciences de Monastir, Université de Monastir, Monatir 5019, Tunisia; Laboratoire de Photonique et de Nanostructures (LPN), UPR20-CNRS, Route de Nozay, Marcoussis 91460, France.

Hamna F. Haneef Wright Center for Photovoltaics Innovation & Commercialization and Department of Physics & Astronomy, University of Toledo, Toledo, OH 43606, USA.

Di Huang Key Laboratory of Luminescence and Optical Information, Beijing Jiaotong University, Ministry of Education, Beijing 100044, China; Institute of Optoelectronic Technology, Beijing Jiaotong University, Beijing 100044, China.

Bouraoui Ilahi King Saud University, Department of Physics and Astronomy, College of Sciences, P.O. 2455, Riyadh 11451, Kingdom of Saudi Arabia; Laboratoire Micro-Optoélectroniques et Nanostructures, Faculté des Sciences de Monastir, Université de Monastir, Monatir 5019, Tunisia.

Maxwell M. Junda Wright Center for Photovoltaics Innovation & Commercialization and Department of Physics & Astronomy, University of Toledo, Toledo, OH 43606, USA.

Laxmi Karki Gautam Wright Center for Photovoltaics Innovation & Commercialization and Department of Physics & Astronomy, University of Toledo, Toledo, OH 43606, USA.

Shou-Yi Kuo Department of Electronic Engineering, Chang Gung University, 259 Wen-Hwa 1st Road, Tao-Yuan 33302, Taiwan; Department of Green Technology Research Center, Chang Gung University, 259 Wen-Hwa 1st Road, Tao-Yuan 33302, Taiwan.

Fang-I Lai Department of Photonics Engineering, Yuan-Ze University, 135 Yuan-Tung Road, Chung-Li 32003, Taiwan; Advanced Optoelectronic Technology Center, National Cheng-Kung University, Tainan 70101, Taiwan.

Xinghua Li Center for Advanced Optoelectronic Functional Materials Research, and Key Laboratory of UV-Emitting Materials and Technology, Northeast Normal University, Ministry of Education, 5268 Renmin Street, Changchun 130024, China.

Yichun Liu Center for Advanced Optoelectronic Functional Materials Research, and Key Laboratory of UV-Emitting Materials and Technology, Northeast Normal University, Ministry of Education, 5268 Renmin Street, Changchun 130024, China.

Na Lu Center for Advanced Optoelectronic Functional Materials Research, and Key Laboratory of UV-Emitting Materials and Technology, Northeast Normal University, Ministry of Education, 5268 Renmin Street, Changchun 130024, China.

Shiqiang Luo School of Energy and Environment, City University of Hong Kong, Tat Chee Avenue, Hong Kong, China.

Ridha M'ghaieth Laboratoire Micro-Optoélectroniques et Nanostructures, Faculté des Sciences de Monastir, Université de Monastir, Monatir 5019, Tunisia.

Hassen Maaref Laboratoire Micro-Optoélectroniques et Nanostructures, Faculté des Sciences de Monastir, Université de Monastir, Monatir 5019, Tunisia.

Claudio Magistris Department of Chemistry and NIS Interdepartmental Centre, University of Torino, Via Giuria 7, I-10125 Torino, Italy.

Murugesan Marikkannan Department of Materials Science, School of Chemistry, Madurai Kamaraj University, Ta mil Nadu, Madurai 625 019, India.

Jeyanthinath Mayandi Department of Materials Science, School of Chemistry, Madurai Kamaraj University, Ta mil Nadu, Madurai 625 019, India.

Kathleen Meehan School of Engineering, University of Glasgow, Glasgow, Scotland G12 8QQ, UK.

Fujun Miao Center for Advanced Optoelectronic Functional Materials Research, and Key Laboratory of UV-Emitting Materials and Technology, Northeast Normal University, Ministry of Education, 5268 Renmin Street, Changchun 130024, China.

Gilles Patriarche Laboratoire de Photonique et de Nanostructures (LPN), UPR20-CNRS, Route de Nozay, Marcoussis 91460, France.

Joshua M. Pearce Department of Materials Science & Engineering, Michigan Technological University, 1400 Townsend, Houghton, MI 49931, USA; Department of Electrical & Computer Engineering, Michigan Technological University, 1400 Townsend, Houghton, MI 49931, USA.

Nikolas J. Podraza Wright Center for Photovoltaics Innovation & Commercialization and Department of Physics & Astronomy, University of Toledo, Toledo, OH 43606, USA.

Pierluigi Quagliotto Department of Chemistry and NIS Interdepartmental Centre, University of Torino, Via Giuria 7, I-10125 Torino, Italy.

Davide Saccone Department of Chemistry and NIS Interdepartmental Centre, University of Torino, Via Giuria 7, I-10125 Torino, Italy.

Mohammed Salah Department of Engineering Mathematics and Physics, Faculty of Engineering, Alexandria University, Elhadara, Alexandria 21544, Egypt; Center of Smart Nanotechnology and Photonics (CSNP), Smart Critical Infrastructure (SmartCI) Research Center, Alexandria University, Elhadara, Alexandria 21544, Egypt.

Effat Samir Center of Smart Nanotechnology and Photonics (CSNP), Smart Critical Infrastructure (SmartCI) Research Center, Alexandria University, Elhadara, Alexandria 21544, Egypt; Department of Electrical Engineering, Faculty of Engineering, Alexandria University, Elhadara, Alexandria 21544, Egypt.

Larbi Sfaxi Laboratoire Micro-Optoélectroniques et Nanostructures, Faculté des Sciences de Monastir, Université de Monastir, Monatir 5019, Tunisia.

Changlu Shao Center for Advanced Optoelectronic Functional Materials Research, and Key Laboratory of UV-Emitting Materials and Technology, Northeast Normal University, Ministry of Education, 5268 Renmin Street, Changchun 130024, China.

Nader Shehata Bradley Department of Electrical and Computer Engineering, Virginia Polytechnic Institute State University, 302 Whittemore Hall, VA 24061, USA; Department of Engineering Mathematics and Physics, Faculty of Engineering, Alexandria University, Elhadara, Alexandria 21544, Egypt; Center of Smart Nanotechnology and Photonics (CSNP), Smart Critical Infrastructure (SmartCI) Research Center, Alexandria University, Elhadara, Alexandria 21544, Egypt.

Yasemin Tezsevin TNO Applied Sciences, High Tech Campus 21, Eindhoven 5656 AE, The Netherlands.

Joop van Deelen TNO Applied Sciences, High Tech Campus 21, Eindhoven 5656 AE, The Netherlands.

Guido Viscardi Department of Chemistry and NIS Interdepartmental Centre, University of Torino, Via Giuria 7, I-10125 Torino, Italy.

Kexin Wang Center for Advanced Optoelectronic Functional Materials Research, and Key Laboratory of UV-Emitting Materials and Technology, Northeast Normal University, Ministry of Education, 5268 Renmin Street, Changchun 130024, China.

Lin Wang Key Laboratory of Luminescence and Optical Information, Beijing Jiaotong University, Ministry of Education, Beijing 100044, China; Institute of Optoelectronic Technology, Beijing Jiaotong University, Beijing 100044, China.

Zheng Xu Key Laboratory of Luminescence and Optical Information, Beijing Jiaotong University, Ministry of Education, Beijing 100044, China; Institute of Optoelectronic Technology, Beijing Jiaotong University, Beijing 100044, China.

Jui-Fu Yang Department of Photonics Engineering, Yuan-Ze University, 135 Yuan-Tung Road, Chung-Li 32003, Taiwan.

Jiao Zhao Key Laboratory of Luminescence and Optical Information, Beijing Jiaotong University, Ministry of Education, Beijing 100044, China; Institute of Optoelectronic Technology, Beijing Jiaotong University, Beijing 100044, China.

Ling Zhao Key Laboratory of Luminescence and Optical Information, Beijing Jiaotong University, Ministry of Education, Beijing 100044, China; Institute of Optoelectronic Technology, Beijing Jiaotong University, Beijing 100044, China.

Suling Zhao Key Laboratory of Luminescence and Optical Information, Beijing Jiaotong University, Ministry of Education, Beijing 100044, China; Institute of Optoelectronic Technology, Beijing Jiaotong University, Beijing 100044, China.

About the Guest Editor

Joshua M. Pearce is cross-appointed as an Associate Professor in the Materials Science & Engineering and the Electrical & Computer Engineering at Michigan Technological University. He runs the Michigan Tech in Open Sustainability Technology (MOST) group, which specializes in solar photovoltaic materials, device physics and systems, as well as energy policy and the development of free and open source hardware for science.

Preface to "Photovoltaic Materials and Electronic Devices"

The solar photovoltaic (PV) market continues to grow rapidly throughout the world [1] offering the promise of enabling humanity to utilize sustainable and renewable solar power technology to run society [2]. As the PV industry has grown, the costs have dropped to the point that with favorable financing terms, it is clear that PV has already obtained and surpassed grid parity in specific locations [3]. Now it not uncommon to have solar power be the less expensive option (lower levelized cost of electricity) for both homeowners and businesses [3]. This is driving a positive feedback loop, where additional growth is expected. The cumulative global market for solar PV is expected to triple by 2020 to almost 700 GW, with annual demand eclipsing 100 GW in 2019 [1]. This growth is accompanied by an explosion of solar jobs [4]. Solar workers have outnumbered coal workers in the U.S. for some time, but now their ranks have swollen to surpass even the oil and gas industry [4,5]. The remarkable and sustained growth of the PV industry may tempt the solar PV scientist to sit back and relax: perhaps with a congratulatory pat on the back for a job well done. However, our work is not complete.

Fossil fuels still make up over 80% of all energy use in the U.S., for example [6], and are still growing worldwide as the resultant climate destabilization. This climate alteration has 'committed to extinction' 15 -37% of species in investigated regions and taxa by 2050 using relatively optimistic mid-range climate-warming scenarios [7]. As the late Professor Smalley has pointed out, our challenge as PV researchers is not to be content with GWs of PV production, but we must obtain terrawatt (TW) levels to eliminate fossil fuel combustion and enable a safe and stable global climate [8]. Meeting these goals by scaling what we have done will not be easy, as others have shown this would place a significant demand on the current and future supply of raw materials (chemical elements) used by those technologies [9].

To meet these needs, we still have much to do to advance the next generation of photovoltaic materials and solar cell devices [10], to further reduce costs to enable more rapid diffusion of solar energy throughout the globe. This book covers some of the materials, modeling, synthesis, and evaluation of new materials and their solar cells, which can help us reach the goal of a sustainable solar-powered future [2].

Joshua M. Pearce
Guest Editor

References

1. GTM Research. Global PV Demand Outlook 2015-2020: Exploring Risk in Downstream Solar Markets. 2015. https://www.greentechmedia.com/research/report/global-pv-demand-outlook-2015-2020

2. Pearce, J.M., 2002. Photovoltaics --a path to sustainable futures. *Futures*, 34(7), pp.663-674.

3. Branker, K., Pathak, M.J.M. and Pearce, J.M., 2011. A review of solar photovoltaic levelized cost of electricity. *Renewable and Sustainable Energy Reviews*, 15(9), pp.4470-4482.

4. Solar Jobs Census. http://www.thesolarfoundation.org/solar-jobs-census/

5. US solar industry now employs more workers than oil and gas, says report http://www.theguardian.com/business/2016/jan/12/us-solar-industry-employees-grows-oil-gas

6. Energy Information Administration, Monthly Energy Review, March 2015, http://www.eia.doe.gov/emeu/mer/pdf/pages/sec1_7.pdf

7. Thomas, C.D., Cameron, A., Green, R.E., Bakkenes, M., Beaumont, L.J., Collingham, Y.C., Erasmus, B.F., De Siqueira, M.F., Grainger, A., Hannah, L. and Hughes, L., 2004. Extinction risk from climate change. *Nature*, 427(6970), pp.145-148.

8. Smalley, R.E., 2005. Future global energy prosperity: the terawatt challenge. *MRS Bulletin*, 30(06), pp.412-417.

9. Vesborg, P.C. and Jaramillo, T.F., 2012. Addressing the terawatt challenge: scalability in the supply of chemical elements for renewable energy. *RSC Advances*, 2(21), pp. 7933-7947.

10. Green, M., 2006. *Third generation photovoltaics: advanced solar energy conversion* (Vol. 12). Springer Science & Business Media.

Integrated Effects of Two Additives on the Enhanced Performance of PTB7:PC$_{71}$BM Polymer Solar Cells

Lin Wang, Suling Zhao, Zheng Xu, Jiao Zhao, Di Huang and Ling Zhao

Abstract: Organic photovoltaics (OPVs) are fabricated with blended active layers of poly [[4,8-bis[(2-ethylhexyl)oxy]benzo[1,2-b:4,5-b']dithiophene-2,6-diyl][3-fluoro-2-[(2-ethylhexyl)carbonyl] thieno[3,4-b]thiophenediyl]]: [6,6]-phenylC71-butyric acid methyl ester (PTB7:PC$_{71}$BM). The active layers are prepared in chlorobenzene (CB) added different additives of 1, 8-Diiodooctane (DIO) and polystyrene (PS) with different concentrations by spin coating. A small addition, 0.5%–5% by weight relative to the BHJ components, of inert high molecular weight PS is used to increase the solution viscosity and film thickness without sacrificing desirable phase separation and structural order. The effects of the PS are studied with respect of photovoltaic parameters such as fill factor, short circuit current density, and power conversion efficiency. Together with DIO, the device with 3.0 v% DIO and 1 wt % PS shows a high power conversion efficiency (PCE) of 8.92% along with an open-circuit voltage (V_{oc}) of 0.76 V, a short-circuit current (J_{sc}) of 16.37 mA/cm^2, and a fill factor (FF) of 71.68%. The absorption and surface morphology of the active layers are investigated by UV-visible spectroscopy, atomic force microscopy (AFM) respectively. The positive effect of DIO and PS additives on the performance of the OPVs is attributed to the increased absorption and the charge carrier transport and collection.

Reprinted from *Materials*. Cite as: Wang, L.; Zhao, S.; Xu, Z.; Zhao, J.; Huang, D.; Zhao, L. Integrated Effects of Two Additives on the Enhanced Performance of PTB7:PC$_{71}$BM Polymer Solar Cells. *Materials* **2016**, *9*, 171.

1. Introduction

Molecular species with well-defined structures [1–3] are being considered as possible substitutions for conjugated polymer counterparts in the fabrication of bulk heterojunction (BHJ) organic photovoltaics (OPVs) [4–6]. High power conversion efficiencies (PCEs) have been achieved in solution-processed molecular solar cells through a combination of chemical design and deposition methods with optimized morphology [7]. Despite the advantages of the structural precision [8] and purity of materials [9–11], some challenges to control the thickness and morphology [12,13] of the active layer decided by the processing conditions are critical to the properties of solar cells, which reasonably influences the light absorption and recombination of carriers [14].

1

A representative example involves the blends comprised of PTB7/PC$_{71}$BM, which is one of the highest-performing systems, based on the addition of small quantities of a high boiling point additive such as diiodooctane (DIO), a kind of commonly used additive [15] to meliorate the morphology of the blend film [16] and promote the phase separation [17]. However, the functions of DIO are still a subject of debate in both polymer and small molecule systems [18], but in the case of PTB7/PC$_{71}$BM it is known to improve the charge transporting by increasing the final crystalline content of the film and allowing the donor phase more than one polymorph during the film formation [19,20]. Nonetheless, adding DIO into the blend film cannot improve the light absorption of the blend film. Even worse, up to now, there is still little research about the negative effect of DIO additives on performance of OPVs [21].

Another effective strategy for PCEs improvement of OPVs is adding a high molecular polystyrene (PS) into the pristine active layer. PS can increase not only increase the pristine solution viscosity but also the film thickness without sacrificing desirable phase separation [7] and structural order and decrease the recombination of electron-hole pairs in the blend film [22,23]. At the same time, PS can improve the light absorption of the blend film. Therefore, in this contribution, polystyrene (PS) was used to fabricate the BHJ polymer solar cell based on PTB7/PC$_{71}$BM as the active layer. DIO and/or PS were varied with different ratios during the solution preparation of the organic active layer. The effect of the PS was investigated in PTB7:PC$_{71}$BM blended films. The morphology of the active layer with different additive ratios has been studied and the related OPVs device performance also has been reported.

2. Experimental Section

2.1. Fabrication of Solar Cells

Devices used materials that were used as purchased. Poly (3,4-ethylenedioxythiophene): poly (styrenesulfonate) (PEDOT:PSS), poly[[4,8-bis[(2-ethylhexyl)oxy]benzo[1,2-b:4,5-b']dithiophene-2,6-diyl][3-fluoro-2-[(2-ethylhexyl) carbonyl]thieno[3,4-b]thiophenediyl]] (PTB7) with a molecular weight of ~200 kg/mol and polydispersity of ~4, [6,6]-phenylC71-butyric acid methyl ester(PC$_{71}$BM), and polystyrene(PS) with a molecular weight of ~370 kg/mol were purchased from Clevios P, 1-Material INC, Nano-C company and Sigma-Aldrich Corporation respectively. PTB7 and PC$_{71}$BM were co-dissolved in chlorobenzene with a weight ratio of 1:1.5 to form the mixed solution with the concentration 20 mg/mL. All organic materials were weighed and dissolved in ambient air conditions.

The devices were fabricated with an architecture of ITO/PEDOT:PSS/PTB7:PC$_{71}$BM/LiF/Al. The indium tin oxide (ITO) glass substrates with a sheet resistance of 10 Ω/Sq were cleaned consecutively in ultrasonic baths containing glass lotion, ethanol and de-ionized water sequentially, and then dried by high pure nitrogen gas. The pre-cleaned ITO substrates were then treated by UV-ozone for 5 min for further cleaning the substrates and improving work function of the ITO substrates. The PEDOT: PSS (purchased from Clevios AI 4083) was spin-coated on the ITO substrates at 3000 rounds per minute (rpm) for 40 s. Then PEDOT: PSS coated ITO substrates were dried in air at 150 °C for 10 min. The substrates were then transferred to a nitrogen-filled glove box (<100 ppm O$_2$ and <0.2 ppm H$_2$O). On the other hand, in order to fabricate the different devices (without DIO additive) as designed in experiment, 0.5%, 1%, 2.5% and 5% of PS by weight were added into PTB7/PC$_{71}$BM mixed solution respectively only half an hour apart, and then the active layers with different ratios of PS were formed by spin-coating on the PEDOT: PSS with same spin-coating parameters, 1 s for acceleration and 120 s with the rotation speed of 1000 rpm. On the top of the active layer, a 0.7 nm interfacial layer LiF was evaporation deposited under 10^{-4} Pa vacuum conditions. The thickness of LiF was monitored by a quartz crystal microbalance. An aluminum cathode layer about 100 nm was then evaporation deposited on LiF layer under 10^{-4} Pa vacuum conditions in same deposition chamber with changed target. The active area was defined by the vertical overlap of ITO anode and Al cathode which is about 4 mm^2. The light mask was not used during I-V measurement, and the potential for edge effects may have an effect on the results. For the convenience of discussion, different films and devices were named and prepared to compare their performances:

Film 1: PTB7:PC$_{71}$BM,
Film 2: PTB7:PC$_{71}$BM, 1 wt % PS
Film 3: PTB7:PC$_{71}$BM, 3 v% DIO
Film 4: PTB7:PC$_{71}$BM, 1 wt % PS and 3 v% DIO
Device 1: ITO/PEDOT:PSS/film1/LiF/Al
Device 2: ITO/PEDOT:PSS/film2/LiF/Al
Device 3: ITO/PEDOT:PSS/film3/LiF/Al
Device 4: ITO/PEDOT:PSS/film4/LiF/Al

2.2. Photovoltaic Characterization

The absorption spectra of films were measured with a Shimadzu UV-3101 PC spectrometer. The thickness of the active layers is measured by an Ambios Technology XP-2 stylus Profiler. The thicknesses of Film 1, Film 2, Film 3 and Film 4 are 85 nm, 110 nm, 73 nm and 102 nm, respectively. The current–voltage (J-V) characteristics of the OPVs were measured using a Keithley 4200 semiconductor characterization system under a simulated AM 1.5G spectrum with power of $100 \ mW/cm^2$ generated by ABET Sun 2000 solar simulator. The corresponding J-V curves were recorded from -1 V to 1 V with an interval of 0.01 V. An incident photon to current conversion efficiency (IPCE) spectrum was measured on Zolix Solar Cell Scan 100. The morphology of the films was investigated by atomic force microscopy (AFM) using a multimode Nanoscope IIIa operated in tapping mode. All the samples were measured with a scan size of $5 \times 5 \ \mu m^2$. The hole mobility and electron mobility of PTB7: $PC_{71}BM$ blend films were measured by space charge limited current (SCLC) method. All the tests were in ambient air conditions.

3. Results and Discussions

The J-V characteristic curves of the OPVs with different PS ratios are shown in Figure 1a. The PV performances of the OPVs are summarized according to the J-V curves and listed in Table 1. Among all the different ratios, it can be found that the device with 1 wt % of PS demonstrates the highest median PCE of 4.56% along with a short-circuit current (J_{sc}) of $10.60 \ mA/cm^2$, an open-circuit voltage (V_{oc}) of 0.79 V, and a fill factor (FF) of 54.50%. The data in Table 1 shows that the PCE improvement is mainly attributed to the enhancement in J_{sc} and FF. To further investigate the mechanism responsible for the enhanced performance of the OPVs with the PS additions, the optimized volume ratio of 1% was used.

It is reported that DIO can improve the morphology of the active layer and enhance the performance of organic solar cells [24]. Consequently, organic solar cells based on PTB7:$PC_{71}BM$ with two additives DIO and PS were prepared to improve photovoltaic properties. The concentration of DIO is 3 wt % according the reference [18], and that of PS is 1 wt % according to the above results. The J-V curves of the OPVs with different additives under illumination of simulated AM1.5G ($100 \ mW/cm^2$) are shown in Figure 1b and summarized in Table 2. Device 1 demonstrates a PCE of 4.11% with a J_{sc} of $10.47 \ mA/cm^2$, a V_{oc} of 0.79 V, and a FF of 49.65%.

As shown in Table 2, with the addition of 1 wt % PS (weight fraction of the BHJ components) in Device 2, J_{sc} increases to $10.60 \ mA/cm^2$ and FF increases to 54.50%, which results in a PCE of 4.56%. If both 3.0 v% DIO and 1 wt % PS are added to the solution prior to spin casting, the PCE of Device 4 is further increased to 8.92 along with a V_{oc} of 0.76 V, a J_{sc} of $16.37 \ mA/cm^2$, and a FF of 71.68%. The improved J_{sc} value is confirmed by measuring EQE (Figure 1c). The maximum EQE value of Device 1 is 43.72% and it is increased to 63.37% for Device 4. The single logarithmic dark current curves show that Device 4, Device 3 and Device 2 have smaller leakage current compared with Device 1, as shown in Figure 1d. It is well-known that the leakage current is determined by the shunt resistance (R_{sh}) [25]. The larger R_{sh}

indicates a lower charge carrier recombination in the active layer. This indicates that DIO and/or PS can effectively restrain the leakage current under reverse bias, which may provide effective charge carrier transport in the blend layers and result in an increase of J_{sc} compared to that of Device 1. The smaller R_s indicates a lower resistance of the semiconductor bulk resistance and a better metal/semiconductor interface connection induced by using additives [25].

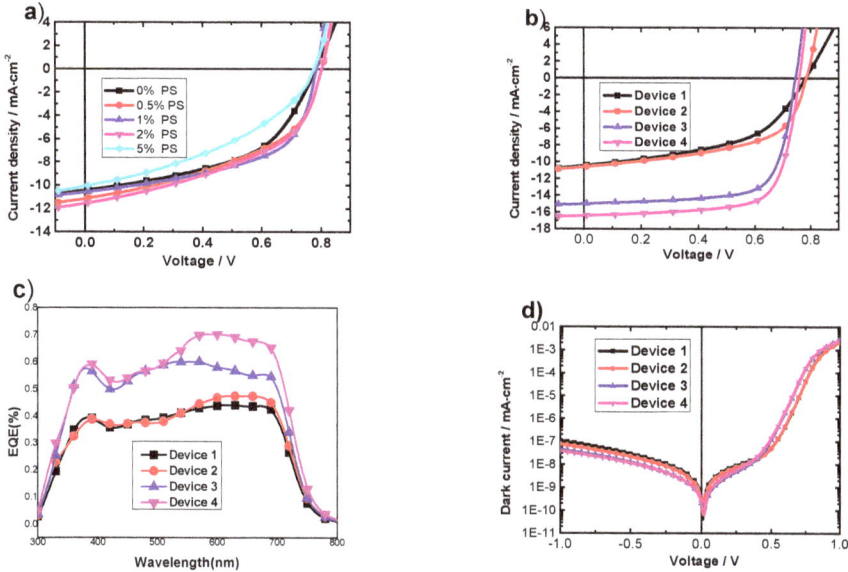

Figure 1. (**a**) The J-V characteristic curves of solar cells with different doping ratios of PS under AM 1.5 light power of 100 mW/cm^2; (**b**) The J-V characteristic curves of solar cells without or with 3 v% DIO and/or 1 wt % PS; (**c**) the external quantum efficiency (EQE) of the devices in the system of PTB7:PC71BM without or with 3 v% DIO and/or 1 wt % PS; (**d**) The J-V characteristics cast from solar cells without or with 3 v% DIO and/or 1 wt % PS in darkness.

Table 1. The PV performance of ITO/PEDOT: PSS/PTB7:PC71BM/LiF/Al photovoltaic devices with different doping ratios of PS.

Doping Ratio	V_{oc} (V)	J_{sc} (mA/cm^2)	FF (%)	PCE (%)
0 wt %	0.79 ± 0.01	10.47 ± 0.09	49.65 ± 0.05	4.11 ± 0.02
0.5 wt %	0.79 ± 0.01	11.15 ± 0.08	46.61 ± 0.06	4.16 ± 0.02
1 wt %	0.79 ± 0.01	10.60 ± 0.08	54.50 ± 0.04	4.56 ± 0.02
2 wt %	0.80 ± 0.01	11.55 ± 0.09	46.68 ± 0.05	4.31 ± 0.02
5 wt %	0.80 ± 0.01	10.07 ± 0.13	38.74 ± 0.08	3.12 ± 0.03

Table 2. The summary of photovoltaic parameters of PTB7:PC71BM system solar cells without or with 3 v% DIO and/or 1 wt % PS.

	V_{oc} (V)	J_{sc} (mA/cm^2)	FF (%)	PCE (%)	Rsh (Ωcm^2)	Rs (Ωcm^2)
Device 1	0.79 ± 0.01	10.47 ± 0.013	49.65 ± 0.02	4.11 ± 0.03	302.8	18.2
Device 2	0.79 ± 0.01	10.60 ± 0.012	54.50 ± 0.09	4.56 ± 0.02	311.5	8.56
Device 3	0.75 ± 0.01	14.23 ± 0.09	71.31 ± 0.05	7.61 ± 0.02	757.6	5.44
Device 4	0.76 ± 0.01	16.37 ± 0.08	71.68 ± 0.04	8.92 ± 0.02	915.8	4.24

Under the same spin-coating condition, the thicknesses of films 1 and 3 are almost the same. When doped with PS, the thicknesses of films 2 and 4 increases and are almost the same. This shows that doping PS encourages an increase the thickness of the active layer without decreasing other electronic properties; for example, Rsh and Rs of corresponding devices doped with PS are improved as shown in Table 2. In order to understand the effect of high-molecular-weight insulating polymers-PS in the mixture of PTB7 and PC$_{71}$BM on the optics properties and the surface morphology, further characterization has been carried out. The absorption spectra of the neat PC$_{71}$BM and PTB7 films are shown in Figure 2a. PC$_{71}$BM has two apparent absorption peaks at 375 nm and 480 nm. The absorption spectra of PTB7 show an apparent complementary absorption in the range from 550 nm to 750 nm. Two broad absorption peaks at around 624 and 682 nm are attributed to the characteristic π-π^* transition of the PTB7 polymer [26,27]. In comparison to the absorption spectra of the films prepared with and without additives, there is no obvious peak shift observed in the film prepared with PS and in the film prepared with pristine CB, as shown in Figure 2b.

It can be found there is no significant difference between Film 1 and Film 3, while there is an increase of obvious relative absorption intensity in the region of 300–800 nm for Film 2 and Film 4. This result indicates that under the same conditions, films with PS additives can harvest solar photons more effectively than the films prepared from CB and CB: DIO. The more light absorbed, the higher photocurrent generated [28]. This is the reason why the J_{sc} increases after the addition of high-molecular-weight insulating polymer PS.

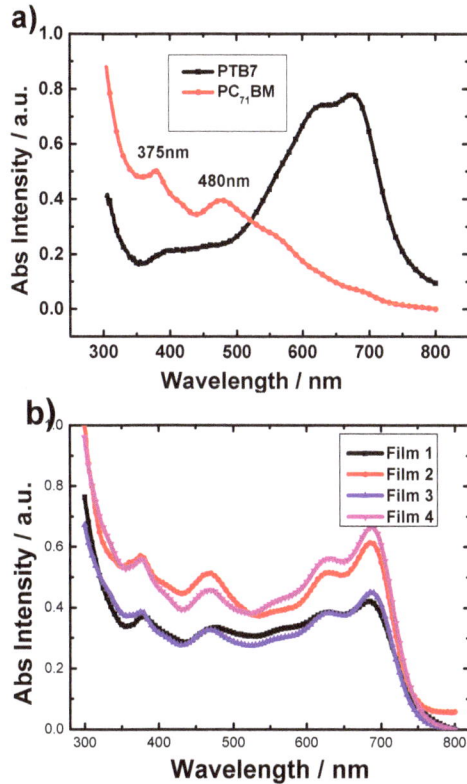

Figure 2. (**a**) The absorption spectra of neat PTB7 and $PC_{71}BM$ films; (**b**) absorption spectra of $PTB7/PC_{71}BM$ films cast from solvents with or without DIO and/or PS additives. The data was normalized by the film thickness.

In order to investigate the effects of PS and DIO on the morphology of the blend films, the surface topography and phase images of the blend films have been studied by atomic force microscopy (AFM) in taping mode (5 μm × 5 μm), as shown in Figure 3. The roughness of Film 1, Film 2, Film 3 and Film 4 are 4.685, 4.855, 1.261 and 1.277 nm, respectively. Obviously, the roughness of the blend films decrease after the addition of DIO, and the phase separation are more finely compared with Film 1 and Film 2, as shown in the phase images. The addition of DIO to the casting solvent results in smaller domains and a more finely interpenetrating BHJ morphology, relative to blend films cast without DIO as shown in the phase diagrams. In particular, Film 4 does not reveal significant increases in roughness at the nanoscale compared to Film 3. Also, the film roughnesses of Film 1 and Film 2 are very similar. All this indicates that incorporating the insulating PS within the photovoltaic layer without negative drawbacks in phase separation.

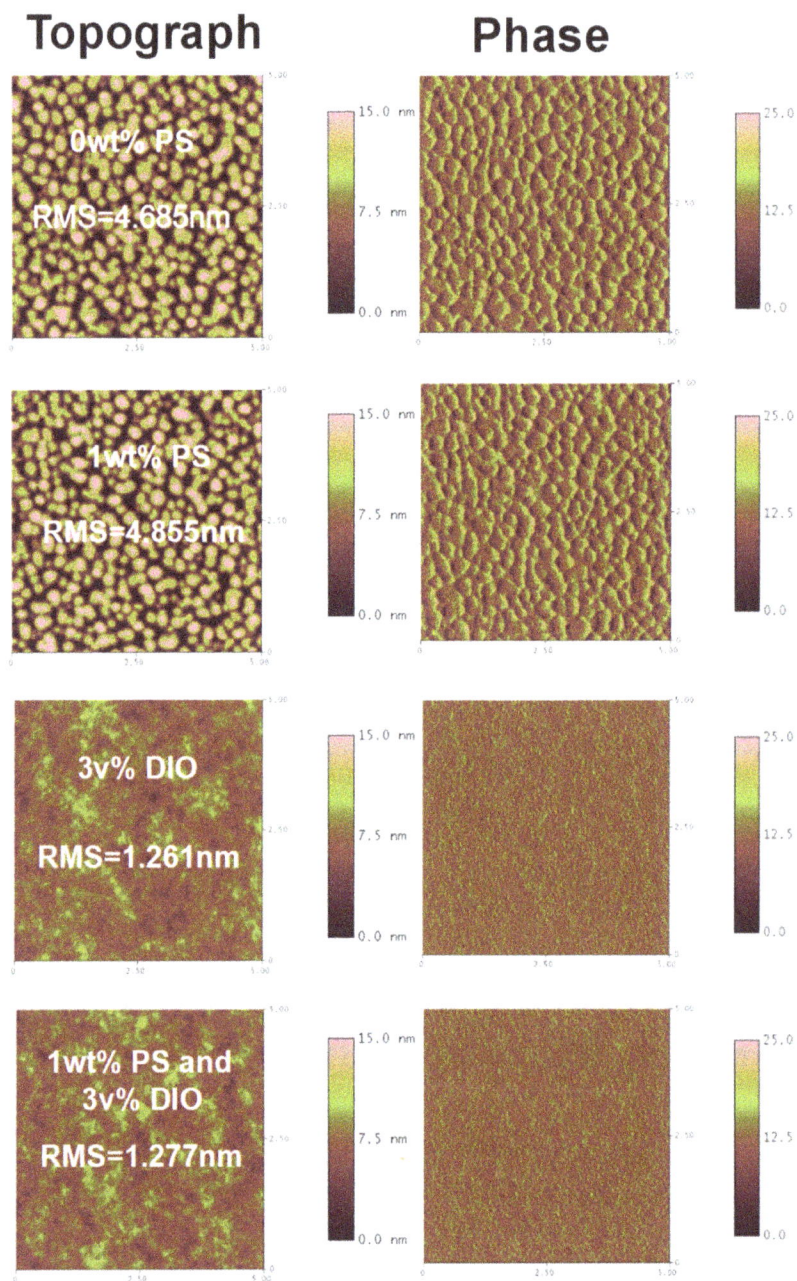

Figure 3. The AFM surface topography and phase images of the PTB7:PC$_{71}$BM films with/without 3 v% DIO and/or 1 wt % PS.

In order to further investigate the effects of additive on the charge carrier transport, the hole-only devices have been fabricated based on pristine PTB7 films and blend PTB7:PC$_{71}$BM films with/without additives, respectively. High-work-function material gold (Au) is used as the cathode to block the back injection of electrons. The dark J-V curves of the hole-only devices with the configuration of ITO/PEDOT:PSS/PTB7/Au and ITO/PEDOT:PSS/PTB7:PC$_{71}$BM/Au are measured and shown in Figure 4. The hole transport through the polymer film is limited due to the accumulation of space charge when a sufficient voltage is applied to this hole-only device. The space charge limited current(SCLC) is described by the equation [29,30]:

$$J = \frac{9}{8}\varepsilon\mu\frac{V^2}{d^3} = \frac{9}{8}\varepsilon_0\varepsilon_r\mu\frac{V^2}{d^3} \tag{1}$$

where ε_0 is the permittivity of free space, ε_r is the dielectric constant of the blend material, μ is the hole mobility, V is the voltage drop across the device and d is the active layer thickness. The parameter ε_r is assumed to be 3, which is a typical value for conjugated polymers. The hole will be collected by the ITO electrode, which is very similar to hole transport process in the OPVs. The J-V characteristics of neat PTB7 and PTB7:PC$_{71}$BM with/without additives of 3 v% DIO and 1 wt % PS fully agree with the SCLC model. According to the J-V curves, the hole current density of the hole-only devices with 1 wt % PS is larger than that of the hole-only devices without any additive, the hole current density of the hole-only devices with 3 v% DIO is larger than that of the hole-only devices with 1 wt % PS and the hole current density of the hole-only devices with additives of both 3 v% DIO and 1 wt % PS is larger than that of the hole-only devices with only 3 v% DIO. This means that hole carrier transport in the hole-only devices with 3 v% DIO and/or 1 wt % PS has been improved compared with that of the hole-only devices without any additive. The result indicates that even though the thickness of the film with PS is greater than that without PS in the same fabricate condition, the hole mobility of the device with PS is better than that without PS, which shows PS can be good for hole carrier transport. Moreover it further demonstrates that the improved hole carrier transport could be one of the reason for the increased J_{sc} of OPVs [30,31].

9

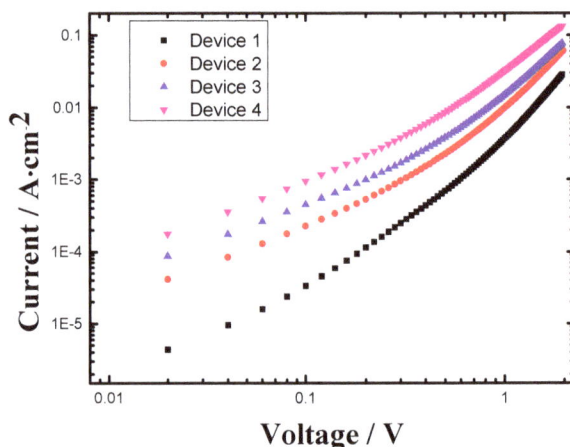

Figure 4. J-V characteristic curves of hole-only devices with/without 3 v% DIO and/or 1 v% PS.

4. Conclusions

A series of OPVs with PTB7:PC$_{71}$BM as the active layer are fabricated to investigate the additive's effects on the performance of the OPVs. DIO and PS are used as the additives. The experimental results of photovoltaic performance reveal an enhancement of J_{sc} from 10.47 to 16.37 mA/cm^2 and FF from 49.65% to 71.68% by adding DIO and PS. As a result, the PCEs of the OPVs are improved from 4.11% to 8.92%, with 117% improvement compared with the OPVs based on PTB7:PC$_{71}$BM without additives. The positive effect of DIO and PS additives on the performance of the OPVs should be attributed to the increased absorption and charge carrier transport and collection.

Acknowledgments: This work was supported by the Fundamental Research Funds for the National Natural Science Foundation of China (No. 61575019, 51272022 and 11474018), the National High Technology Research and Development Program of China (863 Program) (No. 2013AA032205), the Research Fund for the Doctoral Program of Higher Education (No.20120009130005 and 20130009130001), and the Fundamental Research Funds for the Central Universities (No. 2012JBZ001).

Author Contributions: The process design, experimental work and writing of the first draft of the manuscript were all carried out by Lin Wang. Suling Zhao and Zheng Xu supervised every step of the entire work. Jiao Zhao, Di Huang and Ling Zhao collaborated with the AFM analysis.

Conflicts of Interest: The authors declare no conflict of interest.

References

1. Lin, H.W.; Chang, J.H.; Huang, W.C.; Lin, Y.T.; Lin, L.Y.; Lin, F.; Wong, K.T.; Wang, H.F.; Ho, R.M.; Meng, H.F. Highly efficient organic solar cells using a solution-processed active layer with a small molecule donor and pristine fullerene. *J. Mater. Chem. A* **2014**, *2*, 3709–3714.

2. Mishra, A.; Bäuerle, P. Small molecule organic semiconductors on the move: Promises for future solar energy technology. *Angew. Chem. Int. Ed.* **2012**, *51*, 2020–2067.

3. Zhou, J.; Zuo, Y.; Wan, X.; Long, G.; Zhang, Q.; Ni, W.; Liu, Y.; Li, Z.; He, G.; Li, C. Solution-processed and high-performance organic solar cells using small molecules with a benzodithiophene unit. *J. Am. Chem. Soc.* **2013**, *135*, 8484–8487.

4. Cabanetos, C.M.; El Labban, A.; Bartelt, J.A.; Douglas, J.D.; Mateker, W.R.; Fréchet, J.M.; McGehee, M.D.; Beaujuge, P.M. Linear side chains in benzo [1, 2-b: 4, 5-b′ 4dithiophene–thieno [3,4-c] pyrrole-4, 6-dione polymers direct self-assembly and solar cell performance. *J. Am. Chem. Soc.* **2013**, *135*, 4656–4659.

5. Dou, L.; Gao, J.; Richard, E.; You, J.; Chen, C.C.; Cha, K.C.; He, Y.; Li, G.; Yang, Y. Systematic investigation of benzodithiophene-and diketopyrrolopyrrole-based low-bandgap polymers designed for single junction and tandem polymer solar cells. *J. Am. Chem. Soc.* **2012**, *134*, 10071–10079.

6. He, Z.; Zhong, C.; Su, S.; Xu, M.; Wu, H.; Cao, Y. Enhanced power-conversion efficiency in polymer solar cells using an inverted device structure. *Nat. Protoc.* **2012**, *6*, 591–595.

7. Huang, Y.; Wen, W.; Mukherjee, S.; Ade, H.; Kramer, E.J.; Bazan, G.C. High-molecular-weight insulating polymers can improve the performance of molecular solar cells. *Adv. Mater.* **2014**, *26*, 4168–4172.

8. Walker, B.; Kim, C.; Nguyen, T.Q. Small molecule solution-processed bulk heterojunction solar cells†. *Chem. Mater.* **2010**, *23*, 470–482.

9. Chen, W.; Nikiforov, M.P.; Darling, S.B. Morphology characterization in organic and hybrid solar cells. *Energy Environ. Sci.* **2012**, *5*, 8045–8074.

10. Betancur, R.; Romero-Gomez, P.; Martinez-Otero, A.; Elias, X.; Maymó, M.; Martorell, J. Transparent polymer solar cells employing a layered light-trapping architecture. *Nat. Photonics* **2013**, *7*, 995–1000.

11. Dennler, G.; Scharber, M.C.; Brabec, C.J. Polymer-fullerene bulk-heterojunction solar cells. *Adv. Mater.* **2009**, *21*, 1323–1338.

12. Nikiforov, M.P.; Lai, B.; Chen, W.; Chen, S.; Schaller, R.D.; Strzalka, J.; Maser, J.; Darling, S.B. Detection and role of trace impurities in high-performance organic solar cells. *Energy Environ. Sci.* **2013**, *6*, 1513–1520.

13. Guo, S.; Cao, B.; Wang, W.; Moulin, J.F.; Müller-Buschbaum, P. Effect of alcohol treatment on the performance of ptb7:Pc71bm bulk heterojunction solar cells. *ACS Appl. Mater. Interfaces* **2015**, *7*, 4641–4649.

14. Foertig, A.; Kniepert, J.; Gluecker, M.; Brenner, T.; Dyakonov, V.; Neher, D.; Deibel, C. Nongeminate and geminate recombination in ptb7:Pcbm solar cells. *Adv. Funct. Mater.* **2014**, *24*, 1306–1311.

15. Liao, H.C.; Ho, C.C.; Chang, C.Y.; Jao, M.H.; Darling, S.B.; Su, W.F. Additives for morphology control in high-efficiency organic solar cells. *Mater. today* **2013**, *16*, 326–336.

16. Chen, W.; Darling, S.B. Understanding the role of additives in improving the performance of bulk heterojunction organic solar cells. *Microsc. Microanal.* **2015**, *21*, 2439–2440.

17. Guo, S.; Herzig, E.M.; Naumann, A.; Tainter, G.; Perlich, J.; Müller-Buschbaum, P. Influence of solvent and solvent additive on the morphology of ptb7 films probed via x-ray scattering. *J. Phys. Chem. B* **2013**, *118*, 344–350.

18. Lou, S.J.; Szarko, J.M.; Xu, T.; Yu, L.; Marks, T.J.; Chen, L.X. Effects of additives on the morphology of solution phase aggregates formed by active layer components of high-efficiency organic solar cells. *J. Am. Chem. Soc.* **2011**, *133*, 20661–20663.

19. Perez, L.A.; Chou, K.W.; Love, J.A.; van der Poll, T.S.; Smilgies, D.M.; Nguyen, T.Q.; Kramer, E.J.; Amassian, A.; Bazan, G.C. Solvent additive effects on small molecule crystallization in bulk heterojunction solar cells probed during spin casting. *Adv. Mater.* **2013**, *25*, 6380–6384.

20. Kniepert, J.; Lange, I.; Heidbrink, J.; Kurpiers, J.; Brenner, T.J.; Koster, L.J.A.; Neher, D. Effect of solvent additive on generation, recombination, and extraction in ptb7:Pcbm solar cells: A conclusive experimental and numerical simulation study. *J. Phys. Chem. C* **2015**, *119*, 8310–8320.

21. Zhao, L.; Zhao, S.; Xu, Z.; Yang, Q.; Huang, D.; Xu, X. A simple method to adjust the morphology of gradient three-dimensional ptb7-th: Pc 71 bm polymer solar cells. *Nanoscale* **2015**, *7*, 5537–5544.

22. Hsin-Yi, C.; Lan, S.; Yang, P.C.; Lin, S.H.; Sun, J.Y.; Lin, C.F. Poly (3-hexylthiophene): Indene-c60 bisadduct morphology improvement by the use of polyvinylcarbazole as additive. *Sol. Energy Mater. Sol. Cells* **2013**, *113*, 90–95.

23. Bai, Y.; Yu, H.; Zhu, Z.; Jiang, K.; Zhang, T.; Zhao, N.; Yang, S.; Yan, H. High performance inverted structure perovskite solar cells based on a pcbm: Polystyrene blend electron transport layer. *J. Mater. Chem. A* **2015**, *3*, 9098–9102.

24. Zhao, L.; Zhao, S.; Xu, Z.; Gong, W.; Yang, Q.; Fan, X.; Xu, X. Influence of morphology of pcdtbt: Pc71bm on the performance of solar cells. *Appl. Phys. A* **2014**, *114*, 1361–1368.

25. Janssen, R.A.; Nelson, J. Factors limiting device efficiency in organic photovoltaics. *Adv. Mater.* **2013**, *25*, 1847–1858.

26. Liang, Y.; Xu, Z.; Xia, J.; Tsai, S.T.; Wu, Y.; Li, G.; Ray, C.; Yu, L. For the bright future–bulk heterojunction polymer solar cells with power conversion efficiency of 7.4%. *Adv. Mater.* **2010**, *22*, E135–E138.

27. Ochiai, S.; Imamura, S.; Kannappan, S.; Palanisamy, K.; Shin, P.K. Characteristics and the effect of additives on the nanomorphology of ptb7/pc 71 bm composite films. *Curr. Appl. Phys.* **2013**, *13*, S58–S63.

28. Hu, X.; Wang, M.; Huang, F.; Gong, X.; Cao, Y. 23% enhanced efficiency of polymer solar cells processed with 1-chloronaphthalene as the solvent additive. *Synth. Met.* **2013**, *164*, 1–5.

29. Malliaras, G.; Salem, J.; Brock, P.; Scott, C. Electrical characteristics and efficiency of single-layer organic light-emitting diodes. *Phys. Rev. B* **1998**, *58*, R13411.

30. Wang, Z.; Zhang, F.; Li, L.; An, Q.; Wang, J.; Zhang, J. The underlying reason of dio additive on the improvement polymer solar cells performance. *Appl. Surf. Sci.* **2014**, *305*, 221–226.

31. Wang, J.; Zhang, F.; Li, L.; An, Q.; Zhang, J.; Tang, W.; Teng, F. Enhanced performance of polymer solar cells by dipole-assisted hole extraction. *Sol. Energy Mater. Sol. Cells* **2014**, *130*, 15–19.

Terpyridine and Quaterpyridine Complexes as Sensitizers for Photovoltaic Applications

Davide Saccone, Claudio Magistris, Nadia Barbero, Pierluigi Quagliotto, Claudia Barolo and Guido Viscardi

Abstract: Terpyridine and quaterpyridine-based complexes allow wide light harvesting of the solar spectrum. Terpyridines, with respect to bipyridines, allow for achieving metal-complexes with lower band gaps in the metal-to-ligand transition (MLCT), thus providing a better absorption at lower energy wavelengths resulting in an enhancement of the solar light-harvesting ability. Despite the wider absorption of the first tricarboxylate terpyridyl ligand-based complex, Black Dye (BD), dye-sensitized solar cell (DSC) performances are lower if compared with N719 or other optimized bipyridine-based complexes. To further improve BD performances several modifications have been carried out in recent years affecting each component of the complexes: terpyridines have been replaced by quaterpyridines; other metals were used instead of ruthenium, and thiocyanates have been replaced by different pinchers in order to achieve cyclometalated or heteroleptic complexes. The review provides a summary on design strategies, main synthetic routes, optical and photovoltaic properties of terpyridine and quaterpyridine ligands applied to photovoltaic, and focuses on n-type DSCs.

Reprinted from *Materials*. Cite as: Saccone, D.; Magistris, C.; Barbero, N.; Quagliotto, P.; Barolo, C.; Viscardi, G. Terpyridine and Quaterpyridine Complexes as Sensitizers for Photovoltaic Applications. *Materials* **2016**, *9*, 137.

1. Introduction

Dye-sensitized solar cells (DSCs) are photoelectrochemical devices able to convert sunlight into electricity [1]. The architecture and operating principles of these devices have already been extensively reviewed in the literature [2–6], and the photosensitizer represents one of the key components of this device. Different kinds of sensitizers [3,4] have been used so far, including Ru complexes [7], porphyrines [5], phtalocyanines, metal-free dyes [6] (including squaraines [8–10], cyanines [11,12], and push-pull dyes [13]).

Since 1997 [14] the interest in 2,2′:6′,2″-terpyridine (tpy) as ligands in organometallic sensitizers for DSC applications has constantly grown and, in the last three years, more than 80 papers and patents concerning this subject were published. Interest on 2,2′:6′,2″:6″,2‴-quaterpyridines (qtpy) is more recent and has resulted in more than 10 papers (Figure 1).

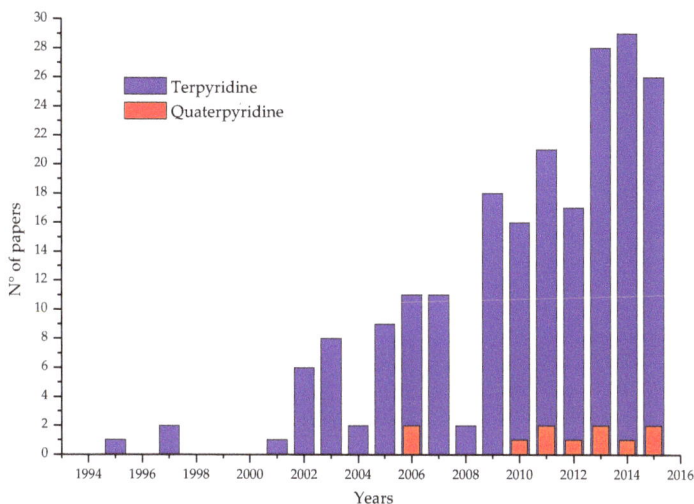

Figure 1. Publications concerning the use of terpyridines (**blue**) and quaterpyridines (**red**) in DSCs. Source: SciFinder (January 2016) [15].

While the general use of polypyridines in Ru complexes sensitizers has already been deeply reviewed in the past by Islam [16], Vougioukalakis [17], and Adeloye [18], or for the electrolytes by Bignozzi *et al.* [19], no insight about the specific structure–properties relationships of tpy and qtpy complexes in the same field have been provided. Thus, we drew our attention on these panchromatic sensitizers with a particular focus on cells performances and device investigation. For this reason works dealing only with computational investigation [20] will not be taken into consideration.

The first use of tpy ligands in DSCs technology was pioneered by Nazeeruddin *et al.* [14], providing good performances owing to their broader absorption with respect to the standard bipyridine-based Ru complexes. The structure proposed in 1997 by the EPFL researchers was named N749 or Black Dye (BD), thanks to its panchromatic absorption (Figure 2, top) and represents a benchmark standard as tpy complex sensitizer. In this dye, ruthenium(II) is complexed by a tpy, the 4,4′,4″-tricarboxy-2,2′:6′,2′-terpyridine (tctpy) and three isothiocyanate ancillary ligands. X-ray diffraction showed a slightly distorted octahedral coordination around the Ru atoms by the three nitrogen donors of tctpy and three nitrogen of isothiocyanate ligands. Very strong intermolecular bonds account for bidimensional arrays, in which the distance between the planes prevents π-stacking between the tpy rings (Figure 2, bottom) [21]. The final BD was prepared by titration with tetrabutylammonium hydroxide in order to deprotonate two of the three carboxylic functions, which proved to be a crucial feature for performances' optimization.

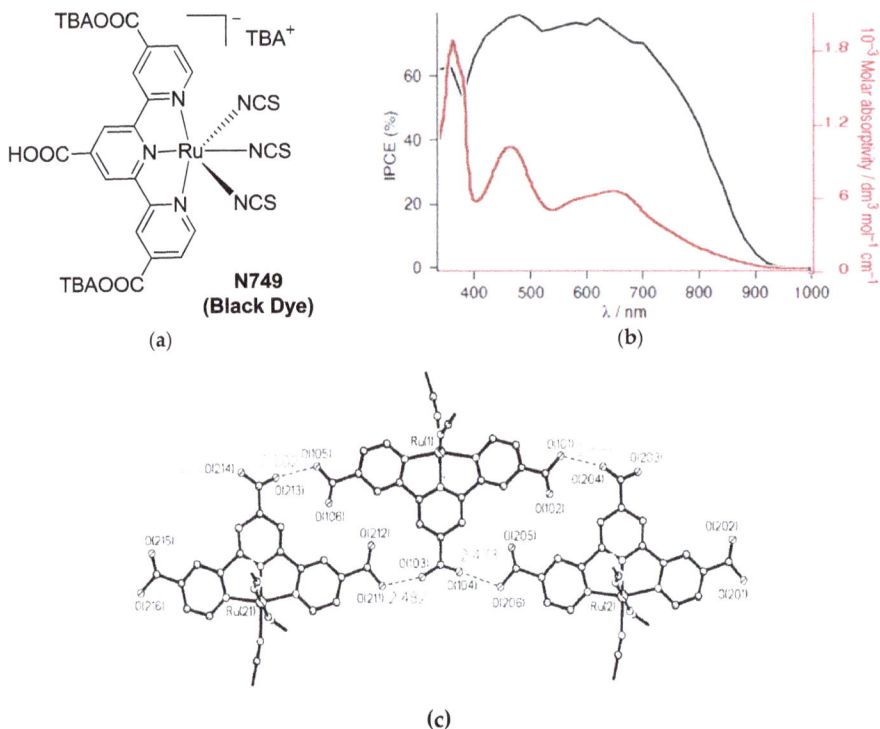

(a)

(b)

(c)

Figure 2. (**a**) Black Dye (BD) or N749 structure; (**b**) light absorption spectrum (**red**) and IPCE (**black**) [12] (Adapted from Ref 12 with permission of The Royal Society of Chemistry); and (**c**) crystal structure showing intermolecular hydrogen bonding [21] (Reprinted with permission from Nazeeruddin, M. K.; Péchy, P.; Renouard, T.; Zakeeruddin, S. M.; Humphry-Baker, R.; Comte, P.; Liska, P.; Cevey, L.; Costa, E.; Shklover, V.; Spiccia, L.; Deacon, G. B.; Bignozzi, C. A.; Grätzel, M. Engineering of efficient panchromatic sensitizers for nanocrystalline TiO$_2$-based solar cells. J. Am. Chem. Soc. 2001, 123, 1613–1624. Copyright 2001 American Chemical Society).

Comparing to bipyridine structures, terpyridines allow to achieve lower band gap for the metal to ligand transition (MLCT), thus providing a better absorption at lower energies and, therefore, broader solar harvesting. The conversion efficiency of BD was first reported as 10.4% (TiO$_2$: 18 μm, dye: 0.2 mM ethanol + 20 mM sodium taurodeoxycholate, electrolyte: 0.6 M DMPII (1,2-dimethyl-3-propylimidazolium iodide), 0.1 M I$_2$, 0.5 M t-bupy (t-butylpyridine), 0.1 M LiI in methoxyacetonitrile) [21], and after further structural tuning (see Section 3.2.5), it was improved up to

11.2% (TiO$_2$: 15 + 7 µm; dye 0.3 mM ethanol / t-butanol 1:1 with 0.6 mM of tetra-butylammonium deoxycholate and 1 mM deoxycholic acid (DCA) as co-adsorbate; electrolyte: 0.6 M DMPII, 0.05 M I$_2$, 0.5 M t-bupy, 0.1 M LiI, 0.1 M GuNCS (guanidinium thiocyanate) in CH$_3$CN) [22]. Despite the wider absorption, performances of BD are not superior to N719 [23] (Figure 3) or other optimized bipyridines complexes [24]. This behavior has been attributed to a lower molar extinction coefficient (7640 M^{-1}·cm^{-1} in DMF) [21] and worse surface coverage of titania [25].

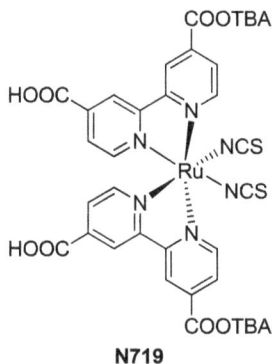

N719

Figure 3. N719 structure.

With the aim to further improve BD performance, several modifications have been carried out concerning each component of the complex. In order to increase the molar extinction coefficient and other features ruthenium was substituted with other metals; thiocyanates were replaced with different pinchers in order to obtain cyclometalated or heteroleptic complexes; and the terpyridine ligand was substituted with a quatertpyridine in order to extend the π-conjugation.

The state of the art of polypyridine structures designed to further improve BD performances is summarized in the next sections. After a survey on the synthetic pathways to obtain tpy and qtpy structures, the three main types of changes underlined before (metal centre, ancillary, and tpy ligands) and their effect on DSCs performances will be taken into account in order to outline a structure-property relationship. Moreover, we remind that DSCs are a complex multivariate system [26], with different components and variables, and that a direct correlation between the photosensitizers' molecular structures and related efficiencies can sometimes lead to inaccurate conclusions. For this reason, we selected literature examples where an internal standard reference (BD, 719 or N3) is reported in order to compare the characteristics of the novel structures. Moreover, specific conditions have been added to selected references.

2. Synthesis

The terpyridine structure was first synthesized in 1932 by Morgan and Burstall [27] as a byproduct of bipyridine synthesis, obtained by dehydrogenation of pyridine in the presence of anhydrous ferric chloride. Nowadays, several synthetic pathways have been developed [28–30], allowing this ligand to reach large applications such as uses in the preparation of Co(II) [31], Os(II) [32], Ru(II) [33] Ir(II) [34,35], Pd(II), Pt(II), and Au(III) complexes [36], supramolecular complexes [37–40], molecular wires [41], polymers [42], in the surface functionalization of nanostructures [43], in the conjugation with amino acids [44], biomacromolecules [45], in the coupling with inorganic nanoparticles [46], and have shown their remarkable activity in other fields such as sensing [47] and catalysis [48,49]. We will report briefly the main strategies used to obtain tpy ligands focusing on the structure–properties relationship in DSCs.

2.1. Terpyridine Core

Tpy structures are mainly prepared through two basic synthetic approaches, which involve either ring assembly or coupling methodologies, as summarized in Scheme 1.

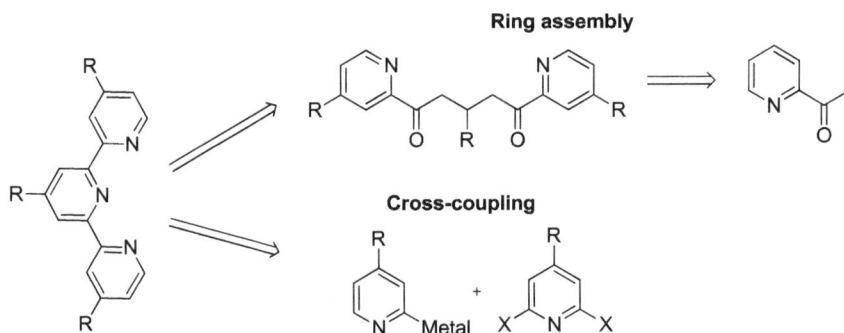

Scheme 1. Retrosynthetic pathways to tpy core.

The first route has been formerly reviewed in 1976 by Kröhnke [50], who reported the synthesis of α,β-unsaturated ketones from 2-acetyl derivatives of pyridine and aldehydes. Then, the intermediate reacts with another 2-acetylpyridine to form a 1,5-diketone that can undergo cyclization to pyridine thanks to ammonia sources such as AcONH$_4$ (Scheme 2). A series of modifications to this procedure has been proposed in order to increase yields or improve the synthetic pathway sustainability [28,51].

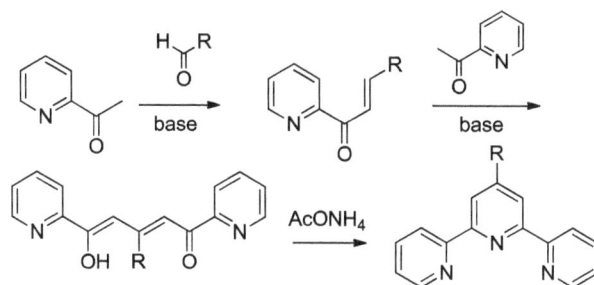

Scheme 2. Example of the Kröhnke pathway.

The second strategy exploits recent advances in organometallic reactions (cross-coupling in Scheme 1). The electron poor pyridines are less effective in the Suzuki reaction [52] due to the weaker electrophilicity of pyridyl-boronates with respect to other organometallic reagents, such as the organo-tin involved in Stille reaction [53].

Noteworthy, the synthetic pathway used to achieve 4,4',4''-tricarboxy-2',6'-terpyridine (tctpy) for Black Dye [54] involves the formation of the terpyridine core starting from 4-ethyl pyridine refluxed with Pd/C over nine days. This procedure was further improved by Dehaudt *et al.* [55]. Among the other possible strategies to obtain a tpy core, it is worth noting an inverse Diels-Alder reaction on 1,2,4-triazine that uses 2,5-norbornadiene as dienophile [56].

2.2. Functionalization of Terpyridines

In order to design complexes suitable for DSCs applications a series of modifications has to be taken into consideration, with the aim of introducing anchoring moieties, donor groups, bulky alkyl chains, or extending the π-conjugation. Cross-coupling reactions represent the most frequently used synthetic tool, while more specific pathways include the formation of carboxylic acid by furan degradation [57–60]. Other common syntheses are dealing with pyridine functionalizations; for example, the pyridine N-oxide is used as an intermediate to obtain halogen and pyrrolidinyl functionalizations [61,62], while 4-pyridones analogues are used to have access to halogens or triflates derivatives [63]. Husson *et al.* reviewed the derivatizations with thienyl [56] and furanyl [64] moieties while recently Woodward *et al.* [65] reported a synthetic strategy to further extend the scope and number of the anchoring moieties on oligopyridines.

2.3. Quaterpyridine Synthesis and Complex Formation

The synthesis and functionalization of qtpy usually exploit the same synthetic strategies used for tpy, namely Kröhnke and coupling reactions. In the latter case N-methyliminodiacetic acid (MIDA [66]) boronates have been successfully applied as key reagents to obtain quaterpyridine ligands in good yields [67] through Suzuki-Miyaura reaction.

In order to obtain Ru(II) complexes of polypyridines, Adeloye *et al.* [18] used Ru *p*-cymene or Ru(III)Cl$_3$ as starting materials and they substituted the chlorines with thiocyanates or other ancillary ligands. Exploiting microwave-assisted synthesis, a facile procedure to obtain a functionalized qtpy ligand and its *trans*-dithiocyanato ruthenium complex has been reported [68] (Scheme 3).

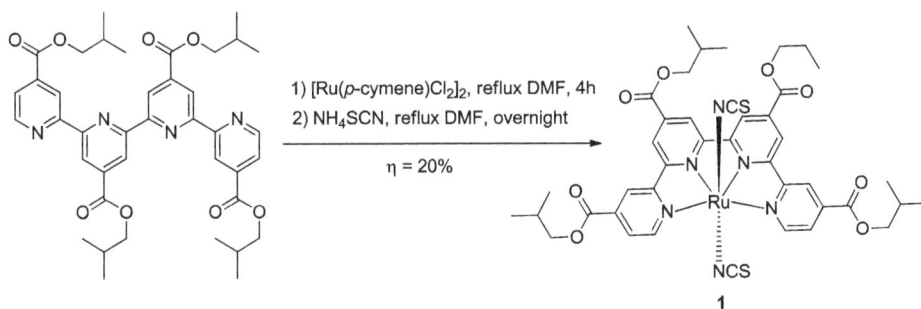

1) [Ru(*p*-cymene)Cl$_2$]$_2$, reflux DMF, 4h
2) NH$_4$SCN, reflux DMF, overnight

$\eta = 20\%$

Scheme 3. Microwave-assisted synthesis of the trans-Ru (II) complex [68].

3. Modifications of Black Dye and Structure-Properties Relationships on Devices

3.1. Terpyridine modification

In this section, tpy based ruthenium complexes bearing three thiocyanates as ancillary ligands will be reviewed, outlining structural modifications on tpy ligand and their effects on DSCs performances.

Molecular engineering on tpy ligands has commonly the aim to extend π-conjugation in order to increase the molar extinction coefficient and further stabilize the LUMO level. In this way more photons can be harnessed and converted thanks to a simultaneous hyperchromic effect and bathochromic shift in the absorption spectra, respectively. Other common structural modifications are the substitution of one of the three pyridines with either a donor group (such as triphenyl amine), in order to enhance the push-pull system character, or a hydrophobic group, in order to reduce recombination with the electrolyte. Particularly interesting are the structural variations related to the anchoring moieties. The tctpy used in BD offers three possible anchoring points, allowing a proper sensitizer-semiconductor coupling and improving the stability of the device. Moreover, alternative anchoring groups, with respect to the carboxylic acid functionality, have been tested. Zakeeruddin [25] proposed a terpyridine functionalized with a phosphonic acid group on 4'-position with the purpose of overcoming the slow desorption of the carboxyl anchoring group from the semiconductor surface in presence of water. Waser [69] proposed a tpy bearing a phosphonic acid functionality, coupled with TiO_2 for DSCs and water splitting applications, while Anthonysamy et al. [70] proposed a 4'-methacryloyloxymethylphenyl moiety as an anchoring group.

As far as the carboxyl anchoring group is concerned, in 2002 Wang et al. [71] tested a 4'-carboxyphenyl substitution (Figure 4), obtaining an appreciable bathochromic shift with respect to N3 (cis-diisothiocyanato-bis(2,2'-bipyridyl-4,4'-dicarboxylic acid) ruthenium(II)), but a sensible loss in short circuit current in comparison with BD occurred, which can be explained by the fewer grafting points on the structure.

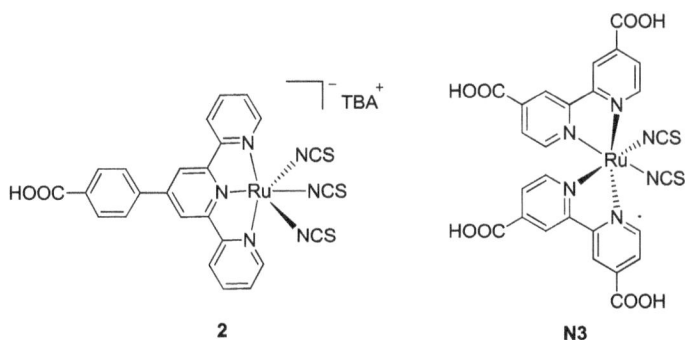

Figure 4. Structure proposed by Wang et al. and N3 dye [71].

Funaki et al. [72] proposed a similar substitution, in which phenylene ethylene moieties (**3a** in Figure 5) were introduced between the COOH functionality and the tpy core, obtaining a better charge injection (12.8 mA/cm²) with respect to dye **2** (6.1 mA/cm²), even if a thicker TiO_2 (36 μm vs. 10 μm) and higher light intensity (100 mW/cm⁻² vs. 78 mW/cm⁻²) were used. The injection efficiency proved to be lower with respect to BD (16.7 mA/cm²), tested in the same conditions. Moreover, when the spacer was represented by two phenylene ethynylene units (**3b** in Figure 4) a higher molar extinction coefficient and slight bathochromic shift were obtained, but a significantly lower J_{sc} value was observed (5.7 mA/cm²) which was ascribed to an increased dye aggregation.

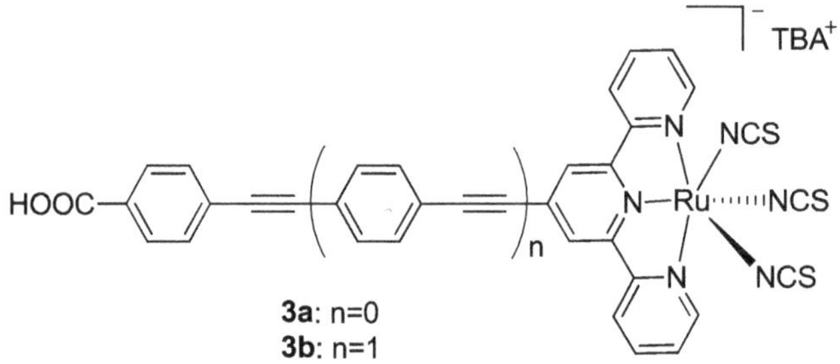

Figure 5. Complexes reported by Funaki *et al.* [72].

McNamara *et al.* [73] reported a ligand similar to **2** bearing a hydroxamic acid instead of the carboxyl moiety. The dye showed promising properties but was not tested on any device.

In 2010, Vougioukalakis *et al.* [74] synthesized a 4′-carboxyterpyridine acid Ru(II) complex (**4a** in Figure 5). With the purpose of increasing the chelating sites, the two outer pyridine rings were also substituted with pyrazine, which resulted in the coordination of a second Ru(II) atom (**4b** in Figure 6).

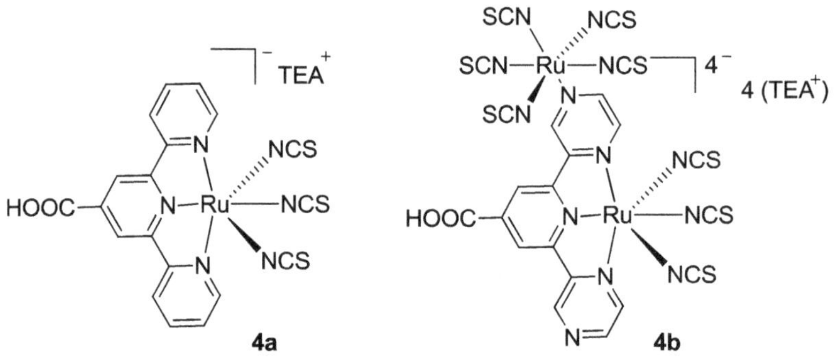

Figure 6. Complexes with one (**4a**) or two (**4b**) metal centers [74].

The overall performances were worse with respect to BD, even if a better absorption on TiO_2 was recorded, due to the greater flexibility of the dyes bearing only one anchoring group, which accounts for a higher number of molecules adsorbed on the surface. Complex **4a**, whose structure is similar to dye **2**, showed similar J_{sc} (6.19 mA/cm^2), but its absorption was hypsochromically shifted with respect to BD. The 2,6-dipyrazinylpyridine ligand (complex **4b**) led to overall lowest performances with 0.27 mA/cm^2 charge injection and 0.02% efficiency (TiO_2: 22 μm, dye 0.3 mM ethanol, electrolyte PMII Ionic Salt, Dyesol). Further improvements in the number of chelated Ru(II) atoms have been reported by Manriquez *et al.* [75] in the preparation of supramolecular structures.

Very recently, Kaniyambatti [76] reported a tpy substituted in 4'- with a cyanoacrylic acid moiety via a thiophene bridge (**5** in Figure 7). The modification leads again to a hypsochromic shift in the absorption spectrum coupled with a higher molar extinction coefficient owing to the extended π-conjugation and strong auxochrome resulting from the thiophene moiety.

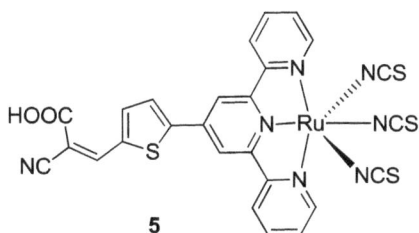

Figure 7. Terpyridine with a cyanoacrylic acid moiety [76].

In 2013, Numata *et al.* [77] proposed a double anchored tpy bearing a 4-methylstyryl substituted in 4"-position (**6** in Figure 8) in order to extend the π-conjugation and to obtain better charge injection with respect to N749. This complex achieved a higher molar extinction coefficient especially on the π-π* transition, and a better IPCE in the same region, which led to an improved efficiency with respect to BD (η = 11.1% ; TiO$_2$: 25 µm; dye: 0.3 mM acetonitrile / *t*-butanol 1:1, 24 h + 20 mM CDCA, electrolyte: 0.05 mM I$_2$, 0.1 M LiI, DMPII, 0.2 M t-bupy in CH$_3$CN).

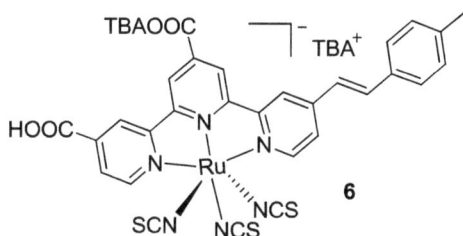

Figure 8. 4-Methylstyryl substituted and double-anchored tpy (HIS-2) [77].

In 2011 Yang *et al.* [78] tested a series of 4,4'-dicarboxy terpyridine bearing a thiophene or a 3,4-ethylenedioxythiophene in 5" position (**7a,b** in Figure 9). The substitution of the latter with a triphenylamino moiety (**7c**) resulted in better performances with respect to BD tested in the same conditions (η = 8.29% *vs.* 6.89%; TiO$_2$: 10 µm + 5 µm, dye: 0.3 mM ethanol + 10 mM chenodeoxycholic acid (CDCA), electrolyte: 0.6 M MDPII, 0.5 M t-bupy, 0.05 M I$_2$, 0.1 M LiI in CH$_3$CN), owing to the higher molar extinction coefficients in the high energy region of the spectrum.

Substitution with hexyl-EDOT (**7b**, EDOT: 3,4-ethylenedioxythiophene) afforded even higher efficiency ($\eta = 10.3\%$ with TiO_2: 15 + 5 μm). Similar modifications have been taken into consideration by Kimura *et al.* [79] (**7d-g** in Figure 9). In the series, structures with hindered hexyloxy-substituted rings resulted in better performances, probably because of the hindrance of alkyl chains towards the electrolyte, thus avoiding the redox couple to interact with titania and considerably reducing the dark current. Among these, the best results were obtained when the electron donor hexyloxy groups on the phenyl ring are in ortho or para positions (**7f** in Figure 9).

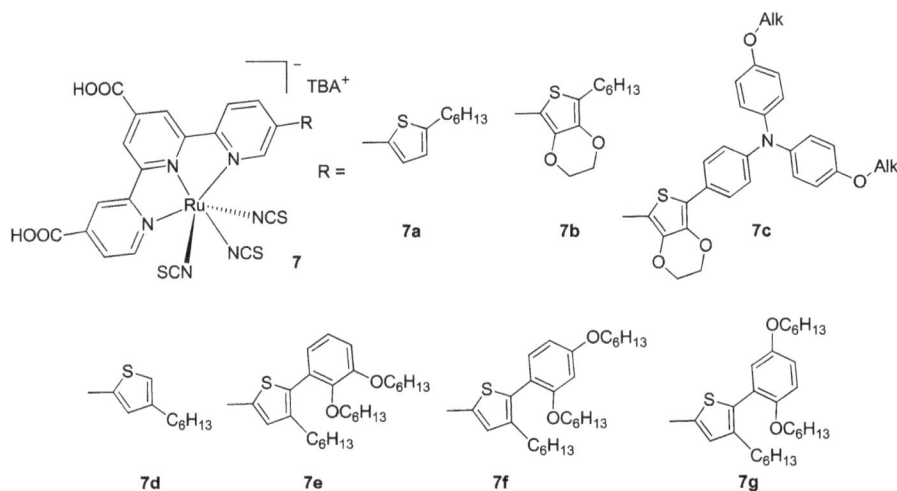

Figure 9. Series of 5″-substituted tpy proposed by Yang (**7a-c**) [78]; and Kimura (**7d-g**) [79].

Very recently, Dehaudt [80] and Koyyada [81] proposed a simple synthetic pathway to achieve 4′-substituted Black Dye analogs (Figure 10) using octylthiophene (**8b**) and hexyl bithiophene (**8d**), pyrrole (**8c**), triphenylamine (**8e**), *t*-butyl phenyl (**8f**), phenoxazine, and phenothiazine groups. While these modifications did not allow to achieve better results respect to the BD in terms of efficiency, they gave an insight into the structure-property relationships, as well as fundamental issues about charge transfer, polarization, or binding. Thienyl-substituted analogues showed better performances with respect to triphenylamino donors, giving an efficiency of 5.57% (TiO_2: 14 + 3 μm, dye: 0.5 mM ethanol / *t*-butanol + 10 mM CDCA, electrolyte: 0.5 M DMPII, 0.5 M t-bupy, 0.1 M LiI, 0.05 M I_2 in CH_3CN).

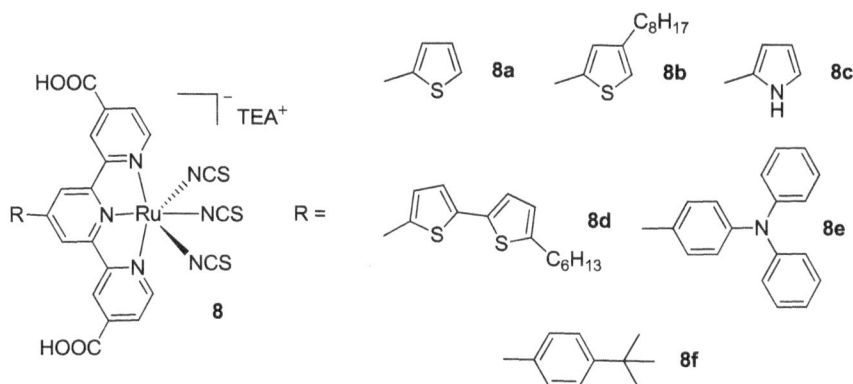

Figure 10. 4′ substituted Black Dye analogs [80].

Ozawa *et al.* proposed a series of tpy having anchoring groups either in the classical 4-, 4′- and 4″-positions or 3′-, 4′-positions, obtaining mono, bis, tri, and tetra-anchored complexes (Figure 11) [82,83]. Substitution with hexylthiophene in 3- or 4-positions was also investigated by impedance spectroscopy (EIS) and open circuit voltage decay (OCVD), revealing that charge recombination with electrolyte solution is largely promoted when compared to the carboxylic-modified one (Figure 10) [84,85]. Efficiencies close to the BD reference were recorded for the tetra-anchored complex **13**, and for the 4″-thienyl dicarboxy substituted complexes **9**. The symmetric substitution with two hexyltiophene groups was also taken into consideration [86,87].

Quaterpyridine Ligand

Tpy modification included the design of tetrapyridines as tetradentate ligands, that were proposed in order to avoid the geometrical isomerism of bipyridine complexes that leads to *cis* and *trans* conformers, showing different optical properties [88]. In fact, *trans* isomers of bipyridines complexes show better photophysical properties, but they are converted by thermal and photoinduced isomerization to the more stable *cis* isomers that, unfortunately, show worse panchromatic absorption. Tetradentate ligands, owing to their planar structure, coordinate the ruthenium in the plane and only leave apical position available for ancillary ligands, thus avoiding the isomerization and ensuring better solar harvesting features. The first example of a tetradentate ligand for DSCs applications was proposed in 2001 by Renouard *et al.* [89] who synthesized a 6,6′-bis-benzimidazol-2-yl-2,2′-bipyridine and a 2,2′:6′,2″:6″,2‴-quaterpyridine bearing ethyl ester functionalities. The qtpy ligand was then characterized for DSCs applications as a complex with Ruthenium (**15**, Figure 12) [90]. The ester moieties showed poor adsorption on TiO$_2$; thus, a further hydrolysis step proved mandatory in order to anchor the dye to the semiconductor surface. Thiocyanate ancillary ligands resulted in blue shifted absorption with respect to chlorine ones due to the stronger σ-acceptor properties of SCN. Remarkable conversion efficiency was recorded, up to 940 nm with 75% IPCE in the plateau region and 18 mA/cm^2 J_{sc} (TiO$_2$: 12 μm, dye: 0.3 mM ethanol / DMSO 95:5, electrolyte: 0.6 M DMPII, 0.1 M I$_2$, 0.5 M t-bupy, 0.1 M LiI in methoxyacetonitrile).

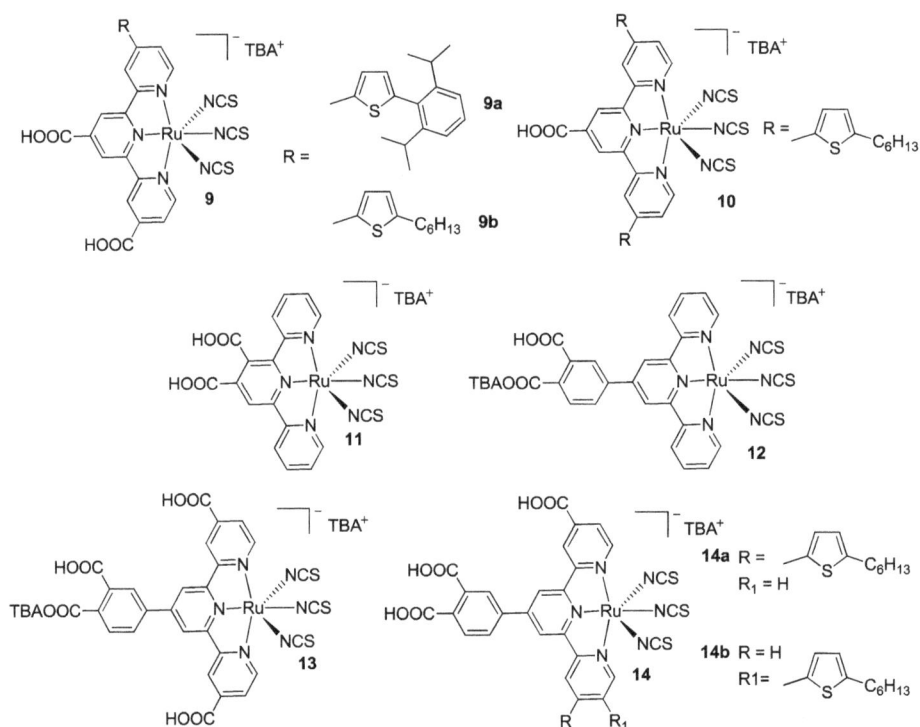

Figure 11. Structures proposed by Ozawa et al. [82–87].

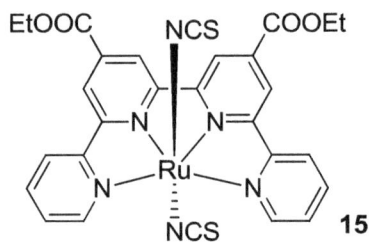

Figure 12. The first qtpy complex applied in DSCs by Renouard et al. [90].

26

A further investigation was reported by Barolo *et al.* [91], in 2006, with the lateral functionalization of the quaterpyridines with *t*-butyl moieties as electron-releasing, bulky groups (**16**, Figure 13). The proposed dye, named **N886**, showed remarkable differences between protonated and non-protonated forms. Wider absorption with respect to N719 was reported, together with a lower molar extinction coefficient and unfavourable alignment of its excited state (as demonstrated by DFT calculations). With the purpose of overcoming these drawbacks, in 2011 the same research group proposed to substitute *t*-butyls with EDOT-vinylene groups, to further extend the π-conjugation (**N1033**, Figure 13) [92]. This complex showed a lower energy gap and a broad IPCE curve having still 33% conversion at 800 nm. The poorer efficiency with respect to **N886** was ascribed to a lower driving force for electron injection, that limits the open circuit potential. The same drawback was also reported for a qtpy substituted with four COOH anchoring moieties (**18**, Figure 13) [68] but its high charge injection and an optimization of the electrolyte composition led to a record efficiency for qtpy Ru-complexes of 6.53% (TiO$_2$: 12 + 5 μm, dye: 0.18 mM *t*-butanol / CH$_3$CN 1:1 with 10% DMF, electrolyte: 1.0 M dimethylimidazolium iodide, 0.03 M I$_2$, 0.1M CDCA, 0.1M GuSCN, 0.23 M LiI in valeronitrile / CH$_3$CN 15:85). Co-sensitization with D35, in order to enhance conversion at higher frequencies, was also reported.

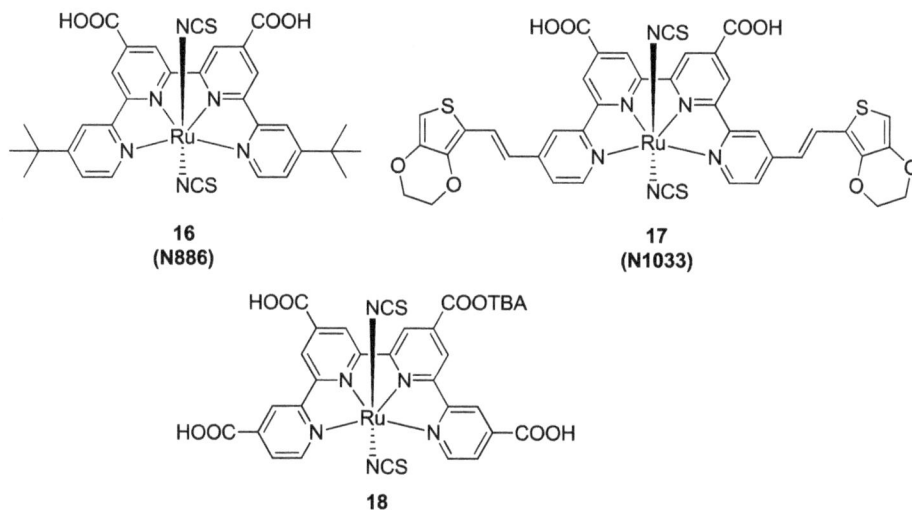

Figure 13. Qtpy complexes investigated by Barolo *et al.* [68,91,92].

3.2. Substitution of Ancillary Ligands: Heteroleptic and Cyclometalated Complexes

A further modification on terpyridine complexes involved the substitution of commonly used thiocyanate ligands with other ancillary ligands. The monodentate thiocyanate ligand has the role to tune the spectral and redox properties of the sensitizers acting on the destabilization of the metal t_2g orbital [93]. By exchanging these ligands with σ-donor groups, it was possible to tune the photochemical properties of the complex, and to minimize the drawbacks associated with these monoanchored ligands. In fact, the possible formation of isomers, owing the bidentate character of the thiocyanate ligand causes a decrease in the synthetic yield [21,78,94]. Moreover the weak Ru-NCS bond itself leads to a decreased stability of the complex and, more importantly, thiocyanate lacks of an effective chromophore that could improve IPCE, particularly at shorter wavelengths. All these features encouraged the engineering of new heteroleptic cyclometalated complexes starting from Black Dye, by exchanging one or more thiocyanate ligands. A drawback affecting this kind of modification is the destabilisation of HOMO orbitals that can lead to a lower driving force in the dye regeneration by the electrolyte.

Strategies for the design of Ru tridentate heterocyclic ligands tailored to tune the properties of the excited state were recently reviewed by Pal *et al.* [95]. Medlycott [96] in 2005 surveyed the strategies for improving the photophysical properties of tridentate ligands commonly considered weaker than bipyridine ones, and Hammarstrom *et al.*, in 2010 [97], investigated the possibility to expand their bite angle. In the following paragraphs we will report an overview of ancillary ligands properly synthesized to tune the photoelectrochemical properties of tpy for applications in DSCs.

3.2.1. Bipyridines

Ancillary ligand exchange was pioneered in 1997 by Zakeeruddin *et al.* [25] who substituted two of the three thiocyanates with a 4,4'-dimethyl-2,2'-bipyridine. In this case, the tpy ligand was not represented by tctpy, but by a simpler tpy with a phosphonic acid anchoring group (Figure 14).

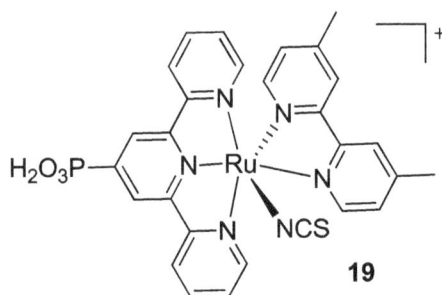

Figure 14. First example of tpy Ru-complex showing a bipyridine instead of two thiocyanates [25].

This research topic became of interest again when, in 2011, Chandrasekharam *et al.* [98] proposed to substitute two thiocyanate ancillary ligands with a bipyridine having electron donor styryl moieties in 4,4'- position (**20a,b**, Figure 15). Worse panchromatic behavior was observed with respect to BD, but also better performances in device, owing to an increased molar extinction coefficient in the visible region. A low value of fill factor led to a 3.36% best efficiency, higher with respect to that of BD evaluated in the same conditions (TiO$_2$: 9 + 4.8 μm, ethanol solution, Z580 electrolyte: 0.2 M I$_2$, 0.5 M GuSCN, 0.5 M N-methylbenzimidazole in [bmim] [I] / 1-ethyl-3-methylimidazolium tetracyanoborate 65:35). Similar bipyridines, slightly modified in the styryl substitution, were also tested by Giribabu *et al.* [99] (**20c**, Figure 15). A more positive oxidation potential with respect to BD under the same conditions has been reported (0.78V *vs.* 0.60V) which was associated with a more negative reduction (−1.30V *vs.* −1.10V) explaining the loss in panchromatic absorption.

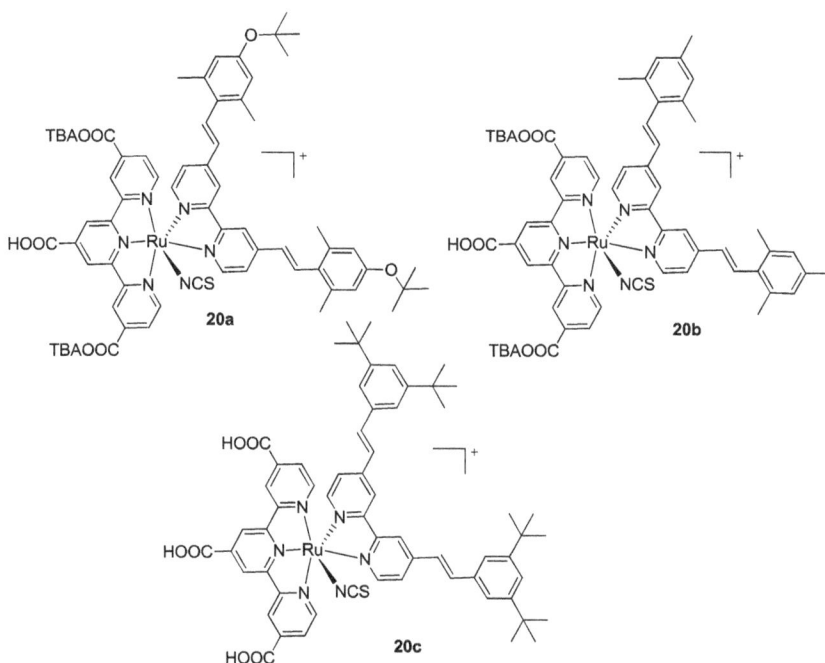

Figure 15. Monothiocyanate complexes proposed by Chandrasekharam [98] (**top**) and Giribabu [99] (bottom).

Very recently, Koyyada *et al.* [100] reported other bipyridines 4,4'- substituted with fluoren-2-yl (**21a** in Figure 16) or carbazol-3-yl (**21b**) groups, as ancillary ligands. Even if the proposed structures reported good molar extinction coefficients and

29

favourable oxidation and reduction potentials, the overall performances were quite low, mainly due to the poor generated photocurrent that was possibly related to an unfavorable localization of LUMO, far from the anchoring sites on titania.

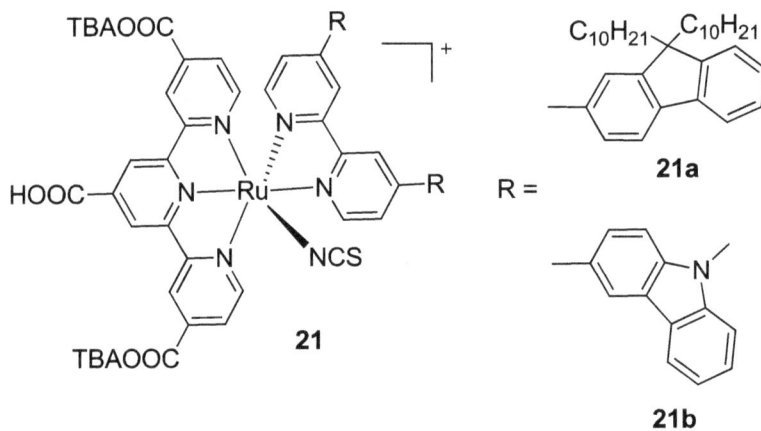

Figure 16. Bipyridine ancillary ligands with fluoren-2-yl or carbazol-3-yl substitutions [100].

In 2015 Pavan Kumar *et al.* [101] modified complex **6** [77] by substituting two thiocyanates with an asymmetrical bipyridine ligand bearing hexylthiophene and mesityl subtituents on each pyridine ring (**22**, Figure 17).

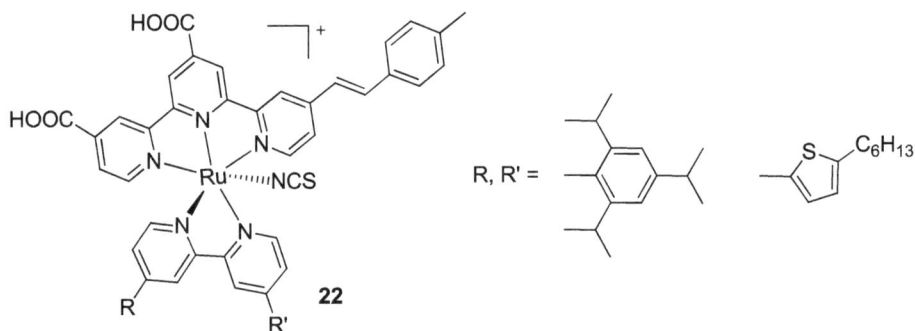

Figure 17. Ancillary ligands modifications of complex **6** [101].

In the same paper, a Ru complex was reported, in which the bipyridine bears two carboxyl substituents. While having four anchoring groups, this complex led to lower efficiencies (**23**, Figure 18). With similar purposes, Kanniyambatti [76] modified complex **5**, achieving a three-anchored sensitizer (**24**, Figure 18) with higher molar extinction coefficient and higher efficiency with respect to both complex **5** and

BD tested in the same conditions (η = 7.5 vs 6.1%; TiO$_2$: 10 + 4 μm, dye: 0.5 mM t-butanol / acetonitrile 1:1 with with CDCA 0.5 mM, electrolyte: 0.6 M [bmim][I], 0.03 M I$_2$, 0.1 M GuSCN and 0.5 M t-bupy in CH$_3$CN / valeronitrile 85:15).

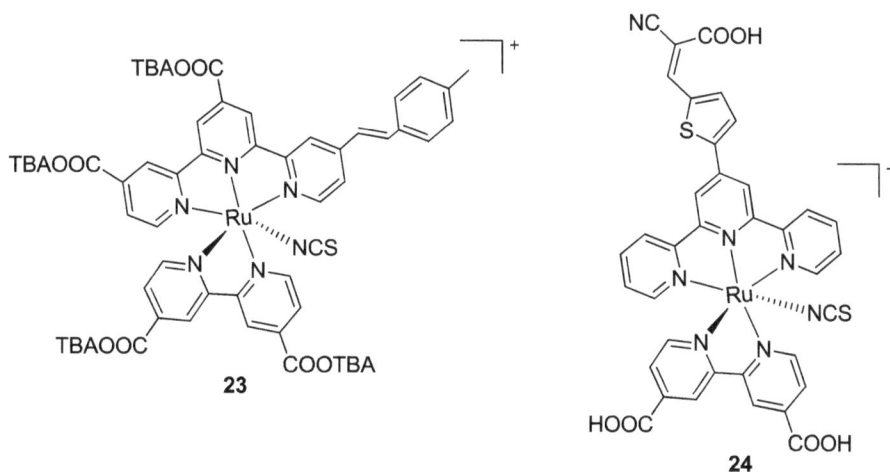

Figure 18. Four (**23**) and three (**24**) anchored complexes by Pavan Kumar [101] and Kanniyambatti [76].

All these modifications were in line with the results from Giribabu, who proposed a Ru complex with 4,4′-dicarboxybipyridine and a tpy ligand bearing the same electron donor in 4,4′,4″- positions (t-butyl or biphenyl amino substituted styryl moieties) (**25a-b**, Figure 19) [102]. In this case, a further enhancement in π-conjugation led to increased molar extinction coefficients and improved performances. Similar complexes that bear donating groups on the terpyridine and electron withdrawing/grafting moieties on a bidentate ligand have been proposed by Mosurkal [103], Erten-Ela [104] and, more recently, by Mongal [105]. In the first case, the anchoring moiety was provided by 4,4′-dicarboxy-2,2′-bipyridine. Mono and dinuclear ruthenium complexes were compared on the device, where the latter one gave better performances. In the second case, the bidentate ligand was represented by a phenantroline substituted with phenyl sulfonic acid moieties in order to graft and sensitize TiO$_2$ and ZnO.

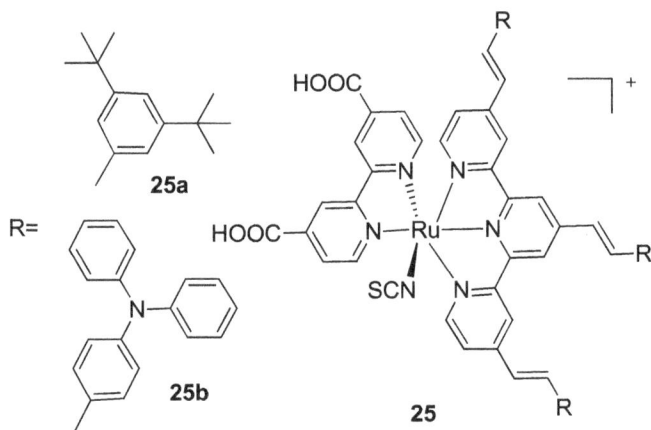

Figure 19. Tpy extended with substituted styryl moieties by Giribabu [102].

3.2.2. Bis-Terpyridine

Stergiopoulos *et al.*, in 2005 [106], replaced all the thiocyanates with another terpyridine. In the resulting heteroleptic complex, one tpy was substituted in 4'- with a *p*-iodophenyl moiety and the other one with a *p*-phenylphosphonic acid, in order to allow the grafting to TiO_2 semiconductor in a solid state device (**26**, Figure 20).

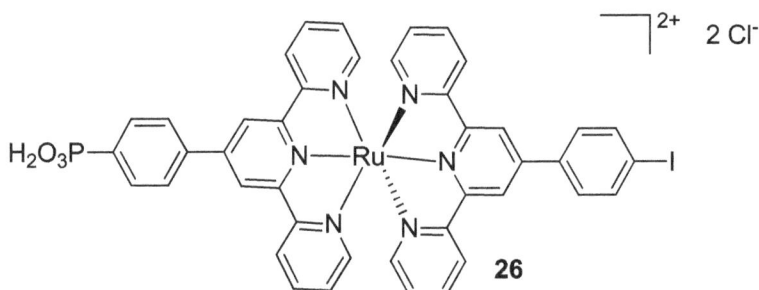

Figure 20. Bis-tpy complex proposed by Stergiopoulos *et al.* [106].

In the same year Houarner *et al.* [107] proposed another bis-tpy complex with a phosphonic acid as the anchoring group on one terpyridine and oligothiophene moieties on the other one, in order to increase the interaction between dye and hole transporting material (**27**, Figure 21). Low performances of this class were attributed to an undesired localisation of the LUMO orbital on thiophenes and, as a consequence, to a difficult charge injection into the TiO_2. In order to improve the performances, the same group in 2007 introduced an unconjugated bridge between the tpy and the polythiophene moiety [108].

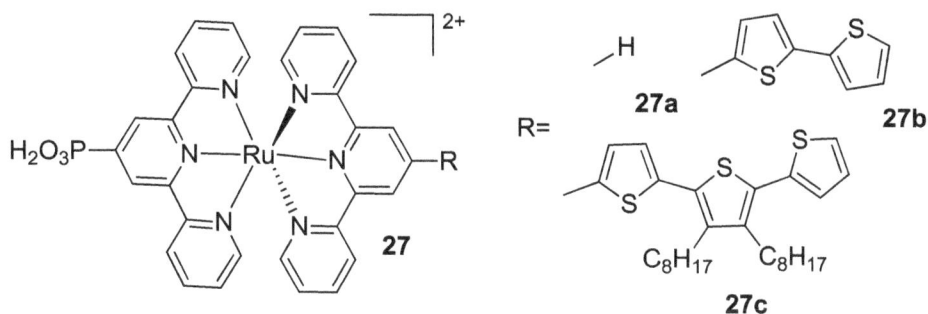

Figure 21. A first series of bis-tpy complexes proposed by Houarner *et al.* [107].

Further improvements to the Houarner series were reported in 2007 [109] by introducing a thiophene π-conjugated bridge between the terpyridine and the phosphonate anchoring group, improving the photoconversion efficiency (**28**, Figure 22). The thiophene spacer proved to be an interesting and efficient relay in the molecular design; however, overall low efficiencies were obtained, owing to a lower driving force for charge injection.

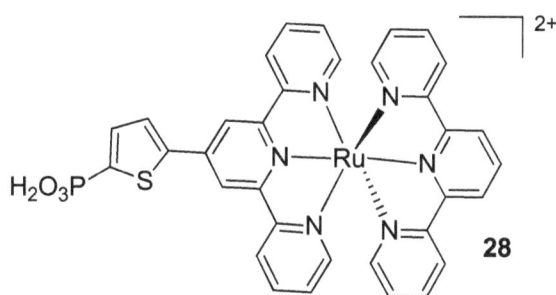

Figure 22. A structural variation of bis-py Ru complex proposed by Houarner *et al.* [109].

Krebs and his research group [110] further investigated bis-tpy Ru complexes using bromophenyl, carboxyphenyl, carboxyl acid [111], and ester moieties in order to compare their anchoring properties. Ester moieties showed weaker absorption to TiO₂ with respect to carboxylic acid and non-symmetric complexes reported efficiencies three times higher with respect to symmetric ones. The same group [112,113] and Chan [114] studied bis-tpy Ru-complexes in conjugated polymers, and their application to polymeric solar cells [112,113]. Tpy-bearing polyphenylene-vinylene and thienyl-fluorene units were exploited in order to incorporate the resulting Ru complexes in the polymer chains; carboxyl acid functionalization of the bipyridine moieties resulted in improved efficiency.

Caramori *et al.* [115], using an heteroleptic thienylterpyridine Ru complex, improved the electron collection efficiency owing to an electrolyte based on the combination of cobalt and iron polypyridine complexes.

Very recently, Koyyada [100] replaced all thiocyanates in the BD structure with a tris (*t*-butyl) tpy, thus maintaining tctpy as the anchoring moiety (**29** in Figure 23). The complex showed good optical properties, with a hypsochromic shift in the visible range of the spectrum and a higher molar extinction coefficient respect to BD, but the overall performances were quite low.

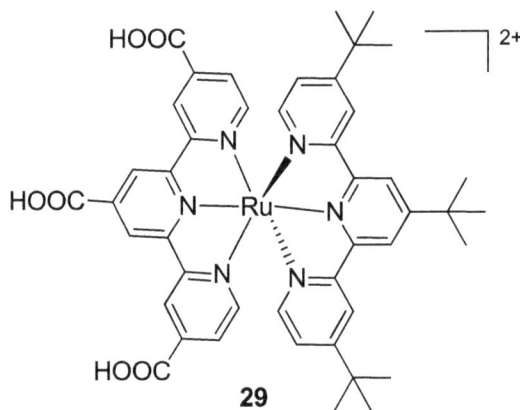

Figure 23. Modification of the BD structure with tris (*t*-butyl) tpy [100].

3.2.3. Phenylpyridine and Pyrimidine

Funaki investigated the possibility to maintain the same terpyridine ligand of Black Dye, tctpy, substituting two thiocyanates with a series of C^N bidentate ligands (**30** in Figure 24) [116–118]. These complexes were designed in order to utilize ancillary ligands with stronger donor properties with respect to thiocyanates in order to destabilize the t_{2g} HOMO orbital, to reduce the band gap and to harness lower energy regions of the solar spectrum. 2-Phenylpyridines as such, and those substituted in 4′ position with a phenyl ethynyl group [118], were used to obtain cyclometalated ruthenium(II) complexes. The wider π-extension allowed to obtain higher molar extinction coefficients and a higher charge injection with an IPCE value of 10% at 900 nm. The main drawback of these complexes was a low oxidation potential that reduced the driving force for dye regeneration. In order to raise the HOMO level and ease the dye regeneration by iodine, the same group [116] extended the C^N ligands series to 2-phenylpyrimidines, substituted on the phenyl ring with trifluoromethyl groups. The CF$_3$ group further reduces the electron donor behavior of the ligand and stabilizes the HOMO level. In this way a 10.7% efficiency was obtained, with respect to 10.1% of BD tested in the same conditions (TiO$_2$: 25 + 6 μm, dye: 0.4 mM ethanol with 40 mM DCA, electrolyte: 0.6 M DMPII, 0.05 M I_2, 0.1 M LiI, 0.5 M t-bupy in CH_3CN). These ligands were further investigated in 2013 [117] by computational studies.

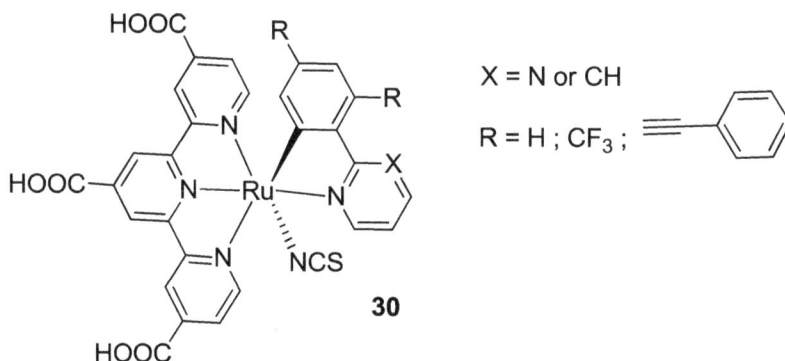

X = N or CH

R = H ; CF_3 ;

30

Figure 24. C^N bidentate ligands proposed by Funaki *et al.* for Ru(II)-complexes [116–118].

3.2.4. β Diketonate Ligands

A series of β-diketonate ligands (**31** in Figure 25) was investigated by Islam *et al.* [119–125] as ancillary ligands alternative to thiocyanates in the BD structure. The strong σ-donating nature of the negatively-charged oxygen donor atom destabilizes the ground-state energy level of the dye compared to BD, leading to a shift of the MLCT transitions to lower energies. In 2002 [125] a Ru(II) complex with 1,1,1-trifluoropentane-2,4-dionato ligand showed efficient panchromatic sensitization of nanocrystalline TiO_2 solar cells. Additionally, a longer alkyl chain (using 1,1,1-trifluoroeicosane-2,4-dionato ligand) [122] prevented surface aggregation of the sensitizer and allowed to avoid or reduce the use of chenodeoxycholic acid. The use of longer alkyl chains may protect the TiO_2 surface, through steric hindrance and hydrophobic effect, preventing the access of electrons to the redox electrolyte, favouring a higher V_{oc}. On the other hand, the bulky alkyl group may not only facilitate the ordered molecular arrangement on the TiO_2 surface, but also keep dye molecules far away each other, thus suppressing intermolecular dye interaction and increasing J_{sc} [126].

In 2006 [123] the same group further modified the β-diketonate ligand with a halogen *p*-chlorophenyl group. Aryl substituents with different electron-donating strength were allowed to control the shift of the low-energy MLCT band and Ru oxidation potential. A very efficient sensitization (η = 9.1%; TiO_2: 20 μm, dye: 0.2 mM CH_3CN / *t*-butanol 1:1 with 20 mM DCA, electrolyte: 0.6 M DMPII, 0.05 M I_2, 0.1 M LiI, 0.07 M t-bupy in CH_3CN), with an IPCE greater than 80% in the whole visible range extending up to 950 nm was obtained. Further substituted β-diketonate ligands were tested in 2011 [119] showing a great potential to tune the photochemical properties.

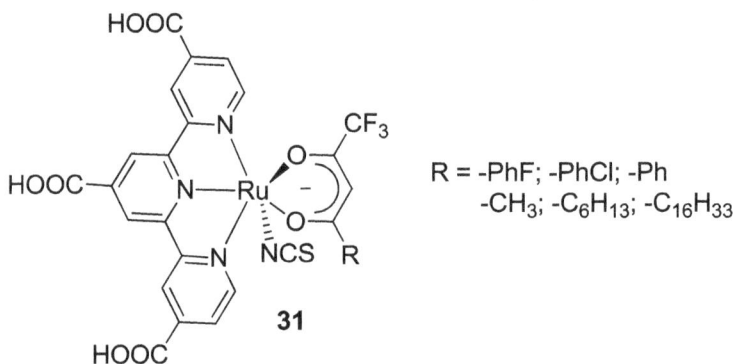

R = -PhF; -PhCl; -Ph
-CH$_3$; -C$_6$H$_{13}$; -C$_{16}$H$_{33}$

31

Figure 25. β-diketonates ligands by Islam *et al.* [119–125].

3.2.5. Pyrazolyl Ligands

Novel N^N bidentate ligands, different from the bipyridines, were proposed by Chen *et al.* [127]. A series of 2-(pyrazol-3-yl)pyridine ligands were used as an alternative to thiocyanate in BD and tested in cells (**32**, Figure 26). These dyes overcome the efficiency of BD tested in the same conditions (η = 10.05 vs 9.07% ; TiO$_2$: 18 + 4 μm, dye: 0.3 mM DMF / *t*-butanol 1:1 with 10 mM DCA, electrolyte: 0.6 M DMPII, 0.1 M I$_2$, 0.1 M LiI, 0.5 M t-bupy in CH$_3$CN) due to their higher molar extinction coefficients between 400 and 550 nm and their extended absorption up to 850 nm, as a consequence of the HOMO destabilization by the pyrazole. The same group reported, in 2011 [128], a series of tridentate 2,6-bis(3-pyrazolyl)pyridine ligands bearing various substitutions in 4- position (**33**, Figure 26). The reported IPCE spectra showed a worse sensitization in the NIR region with respect to N749 but a better conversion in the visible range which accounts for efficiencies up to 10.7% (TiO$_2$: 15 + 5 μm, dye: 0.3 mM ethanol / DMSO 4:1 with 1M CDCA, electrolyte: 0.6 M DMPII, 0.1 M I$_2$, 0.5 M t-bupy, 0.1 M LiI in CH$_3$CN). The results were explained by the bulky ligand effect, which may allow better packing of the dye molecules on the TiO$_2$ surface and prevent interfacial charge recombination. On the other hand, the contribution of the pyridine in the ligand, which is neutral with respect to the negatively charged thiocyanates, might allow the negative dipole moment to be localized closer to the surface, thus affording a higher V_{oc}. Further investigations on these complexes were carried out by replacing the tctpy with a dicarboxytpy ligand substituted in the 5- or 6- position of a terminal pyridyl unit with π-conjugated thiophene pendant chains, obtaining good stability and performances with respect to BD [129]. More recently, the terminal pyridyl unit of the tctpy was replaced with variously substituted quinolines (**34**, Figure 26) reaching good performances (η = 10.19%; TiO$_2$: 15 + 7 μm, dye: 0.3 mM ethanol / *t*-butanol 1:1 with 0.6 mM of tetra-butylammonium deoxycholate, electrolyte: 0.6 M DMPII, 0.05 M LiI, 0.05 M I$_2$, 0.5 M t-bupy in CH$_3$CN) [130]. In this new family of complexes, electron-donating-bulky *t*-butyl substituents on quinoline gave better performances with respect to the electron-withdrawing COOH group. With the *t*-butyl group, in fact, a blueshift for transitions at lower energies was reported together with a hyperchromic effect that improved IPCE and J_{sc}. Further modifications to the bidentate ancillary ligands led to a best result efficiency (η = 11.16%; with the addition of 1mM DCA as co-adsorbate to the dye solution and electrolyte: 0.1 M LiI and 0.1 M GuNCS, 0.5 M t-bupy in CH$_3$CN) [22] when tctpy and hexylthiothienyl-substituted pyrazolyl-pyridine were used to complex ruthenium (II) (**35**, Figure 25). This complex showed worse conversion in the NIR spectral region but improved IPCE in the visible one with respect to BD, thus determining a better efficiency in the same conditions.

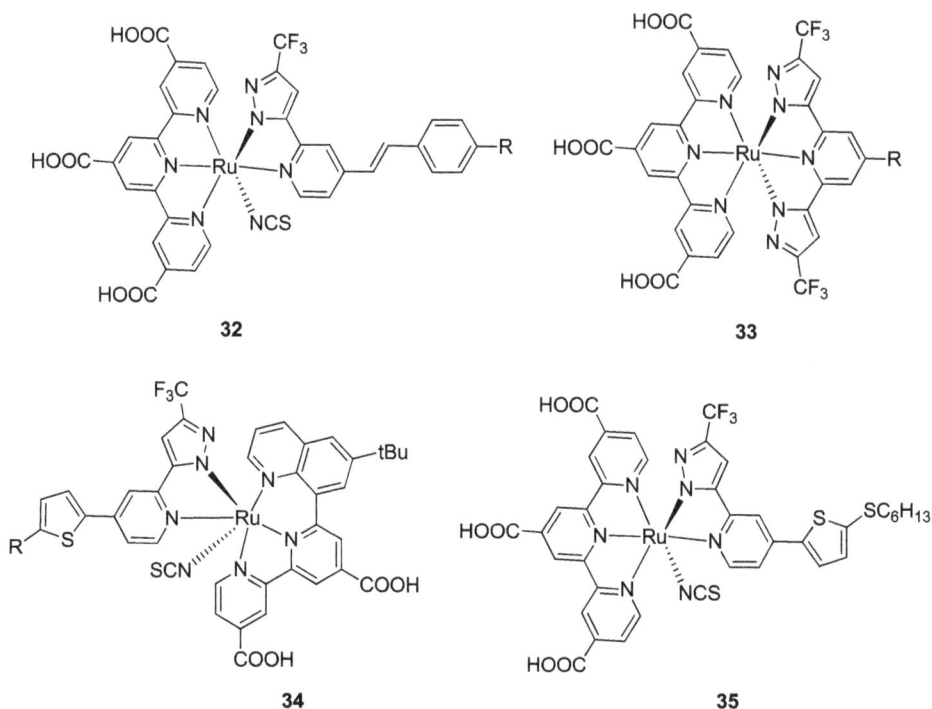

Figure 26. Pyridyl-pyrazolate ligand and quinolyl-bipyridine ligand for Ru(II)-complexes [22,127–129].

Recently, Chang *et al.* [131] reported pyrazolyl-pyridine ancillary ligands bearing a series of donor groups in which the simplest substituents (such as *t*-butyl group) leads to better efficiency with respect to the triphenylamino and benzothiadiazolylgroups.

3.2.6. Phenyl Bipyridines

In 2007 Wadman *et al.* [132] compared a bis-tpy Ru complex bearing one carboxyl group in the 4- position, with two structurally homolog complexes in which the tpy was replaced by 6-phenylbipyridines, with one or two carboxyl groups (**36** and **37**, Figure 27). The N^N'^C Ru(II)-complex with two anchoring groups showed performances similar to N719. Thus, in 2010 [133], the same group further extendend the series, including N^C^N' ligands based on 3,5-bis(2-pyridyl)benzoic acid (**38**, Figure 27).

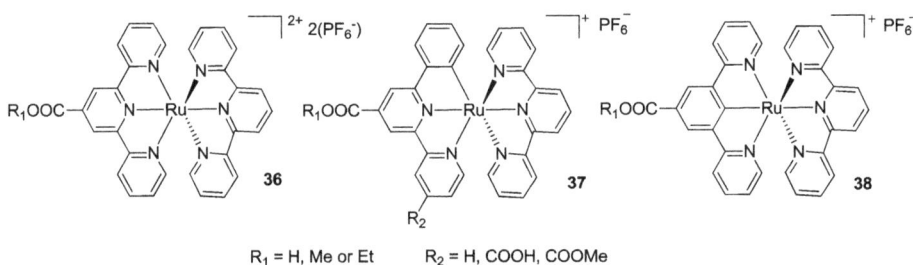

R₁ = H, Me or Et   R₂ = H, COOH, COOMe

Figure 27. Tpy and phenyl-bipyridines complexes investigated by Wadman *et al.* [132].

X-ray structural determination on the mono carboxyl complex (**37**, R_1 = H, R_2 = COOH) showed a distorted octahedral coordination, with the cyclometalated ligand perpendicular to the terpyridine and elongation of the nitrogen to Ru bond opposite to the C-Ru bond. In the solid state, the complex forms dimers via hydrogen bonds between the carboxyl functions (Figure 28).

N^N'^C cyclometalated compounds showed better sensitization properties respect to the bis-tpy complexes; while the lower efficiencies of the N^C^N' complexes were ascribed to a LUMO localization which prevented an efficient electron injection into the TiO_2 conduction band. The replacement of a coordinative Ru-N bond with a covalent carbon-ruthenium bond led to a redshift and to a broadening in the optical absorption of the corresponding ruthenium complex. Functionalization on the N^C^N' ligand with another tpy resulted in the synthesis of dinuclear Ru(II)-complexes [134].

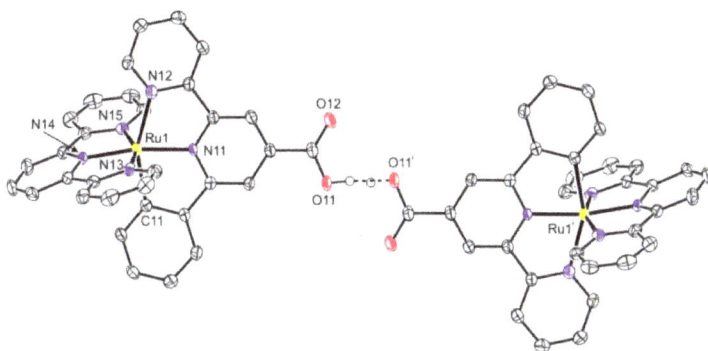

Figure 28. Crystal structure of complex **37** in form of its dimer [132] (Adapted from Ref 131 with permission of The Royal Society of Chemistry).

Kisserwan *et al.* [135] further engineered the 6-phenyl-2,2'-bipyridyl (C^N^N') ligand with a thiophene and carboxylic acid moieties in the 4- and 4'- positions of the bipyridine moiety (**39**, Figure 29). The thienyl group was chosen with the purpose of increasing the molar extinction coefficient, while COOH had the aim to further strengthen the coupling with TiO_2. With respect to Wadman's works, tctpy was used instead of tpy. The work focused more on electrolyte composition than on sensitizer design, providing better performances when CuI was used as an additive. The same group in 2012 [57] extendend the investigation on the 6-phenyl-2,2'-bipyridyl (C^N^N') ligand, studying the influence of either donor or acceptor substituents on the phenyl and the presence of COOH on the bipyridine. When the thienyl group was replaced by COOH, lower efficiencies were observed, attributed to a less efficient electron injection. The best sensitizer was also studied for its long-term stability, showing better results when compared to N719.

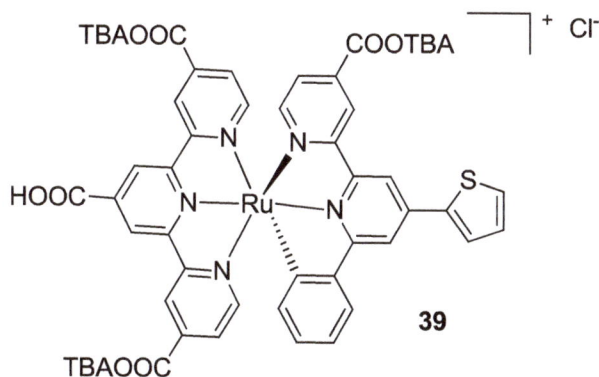

Figure 29. Bis-tpy-based Ru(II) complex proposed by Kisserwan *et al.* [135].

In 2011, Robson *et al.* [136] published an extensive study in which a series of asymmetric bis-tridentated ruthenium complexes was synthesized, whose ligands ranged from terpyridine (N^N'^N") to phenyl-bipyridine (C^N^N') and di-(2-pyridyl)-benzene (N^C^N'), bearing anchoring electron-withdrawing groups on one ligand and, on the other, a thienyl-triphenylamino group as donor counterpart (**40**, Figure 30). A thorough investigation of the photophysical and electrochemical properties was pursued in order to understand the role of the organometallic bond and terminal substituents and to tune the energetic levels. Broad absorption spectra were generated in Ru(II) complexes containing an organometallic bond because of the electronic dissymmetry about the octahedral Ru(II) center. The intensity of the spectra in the visible region was enhanced when the organometallic bond was orthogonal to the principal axis (*i.e.*, C^N^N' ligand). When the anchoring ligand is represented by a N^C^N' tridentate combination, the LUMO is placed remotely from TiO$_2$, and this prevents an efficient charge injection. On the other hand, if the organometallic bond is placed on the donor ligand, HOMO level can be localized either on the triphenyl amino moiety or on Ru(II), maximizing light harvesting in the visible region; while, at the same time, the LUMO on the anchoring ligand ensures an efficient electron transfer towards the semiconductor surface. The highest recorded efficiency reached 8.02% (TiO$_2$: 15 + 4.5 μm, dye: 0.3 mM ethanol, Z1137 electrolyte: 1.0 M 1,3-dimethylimidazolium iodide, 60 mM I$_2$, 0.5 M t-bupy, 0.05 M NaI, 0.1 M GuNCS in CH$_3$CN / valeronitrile 85:15).

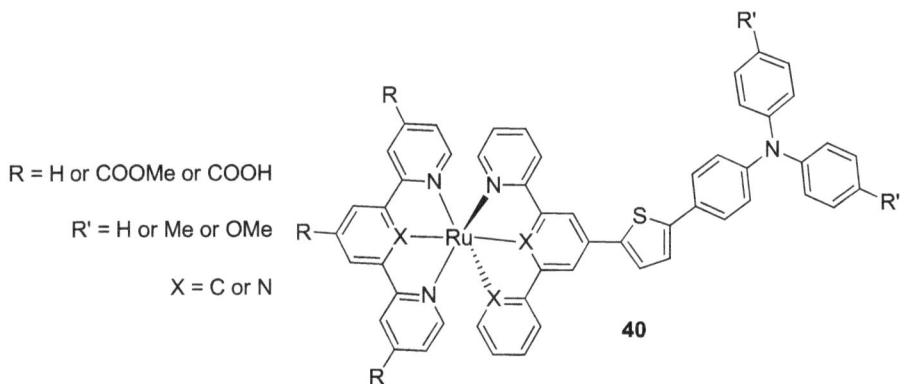

Figure 30. Robson *et al.* [136] series of bis-tridentated ruthenium complexes bearing triphenyl amino groups.

3.2.7. Dipyrazinyl-Pyridine

Another series of bis-tridentate complexes was reported in 2007 by Al-mutlaq *et al.* [137] using dipyrazinyl-pyridine ligands with different substituents on 4'- position, and cathecol moieties as grafting groups (**41**, Figure 31). In comparison to homolog complexes with terpyridine, dipyrazinyl-pyridine led to higher oxidation potential. Exchanging SCN improved HOMO and LUMO while substituting tpy with dipyrazinyl-pyridine lowered these values.

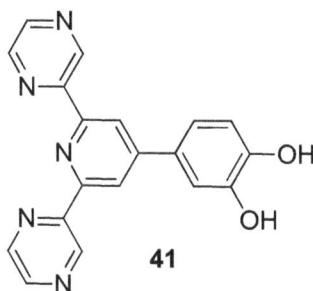

Figure 31. Example of dipyrazinyl-pyridine ligand [135].

Sepehrifard *et al.* [138,139] investigated a series of homoleptic bis-tridentate ruthenium complexes, employing both tpy and dipyrazinyl-pyridine ligands. The poorer performances of the latter ones were attributed to lower LUMO levels and weaker bonding to TiO$_2$. The best results were obtained with terpyridine ligands bearing COOH grafting groups (1.53% efficiency) while the use of dipyrazinyl-pyridine ligands, ester groups or the introduction of a phenylene spacer between the pyridine and the anchoring group all resulted in lower efficiencies.

3.2.8. Triazolate

Schulze *et al.* investigated triazolate as chelating moiety in a series of N^C^N' cyclometalated ligands [140] and N^N'^N" ligands [141]. 1,3-Di(4-triazolyl)benzene and 2,5-di(4-triazolyl)pyridine were used in association with tctpy as the grafting moiety (**42**, Figure 32). In the case of the N^C^N' ligand, the substitution with electron-withdrawing groups such as F or NO$_2$ stabilizes the HOMO energy level providing blueshift and loss in charge injection, while hydrophobic alkyl chains are expected to be beneficial for the long-term stability. The relatively low efficiency obtained as the best result (η = 4.9%; TiO$_2$: 12 + 3 μm, dye: 0.25 mM methanol, electrolyte: 0.6 M 1,3-dimethylimidazolium iodide, 0.06 M I$_2$, 0.1 M LiI, 0.5 M t-bupy, 0.1 M GuSCN in CH$_3$CN) in the case of the N^N'^N" ligand with respect to N749 (6.1% in the same conditions, dipping solution in ethanol) was explained by loss in panchromatic absorption.

Figure 32. Triazolate ligand studied by Schulze *et al.* [140,141].

3.2.9. Other Ligands

C^N^C′ ligands have been tested by Park *et al.* [142] in a series of bis-tridentate ruthenium complexes, exploiting N-heterocyclic carbenes such as 2,6-bis-(3-methylimidazolium-1-yl)pyridine (**43a-c**, Figure 33).

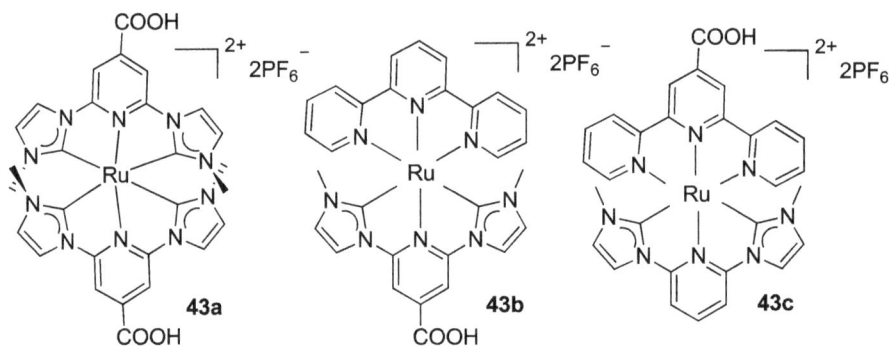

Figure 33. Ru(II) complexes proposed by Park *et al.* [142].

X-ray crystal structure of **43b** shows a typical geometry with both ligands coordinated in a meridional fashion; bond distances between Ru and the coordinated N or C are similar and the carboxyl function is deprotonated (Figure 34). Overall efficiencies were far from N719 tested in the same conditions, a result that was mainly attributed to low charge injection.

Bonacin *et al.* [143] proposed a complex of Ru(II) with carboxyphenyl tpy, thiocyanate, and 8-hydroxy quinoline in order to host a carboxymethyl cyclodextrin anchored to TiO$_2$. Even if poor results were reported (ascribed to high HOMO potential and low regeneration), the host-guest interaction of the dye with the cyclodextrin increased the performances by preventing dye aggregation and limiting the dark current.

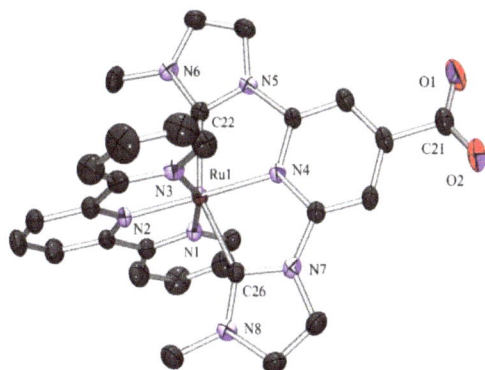

Figure 34. ORTEP drawing of complex **43b** [142] (Reprinted with permission from Park, H.-J.; Kim, K. H.; Choi, S. Y.; Kim, H.-M.; Lee, W. I.; Kang, Y. K.; Chung, Y. K. Unsymmetric Ru(II) Complexes with N-Heterocyclic Carbene and/or Terpyridine Ligands: Synthesis, Characterization, Ground- and Excited-State Electronic Structures and Their Application for DSSC Sensitizers. Inorg. Chem. 2010, 49, 7340–7352. Copyright 2010 American Chemical Society).

Kinoshita *et al.* [144,145] spent efforts in order to further extend the absorption of BD. In conventional Ru(II) complexes, short-lived ^{1}MLCT states immediately relax to long-lived ^{3}MLCT states through intersystem crossing. The spin-forbidden singlet-to-triplet transition from HOMO to ^{3}MLCT has been observed for a phosphine-coordinated Ru(II) sensitizer (**44** in Figure 35), providing light conversion up to 1000 nm and unprecedented charge injection (26.8 mA/cm^{-2}). Unfortunately no evidence about long-term stability of this complex was reported.

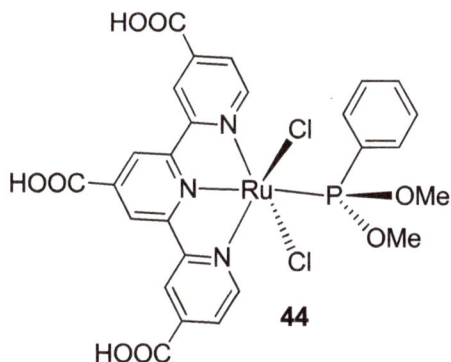

Figure 35. Phosphine-coordinated Ru(II) sensitizer by Kinoshita *et al.* [144].

Recently, Li used 2,2′-dipyrromethanes as N^N′ bidentate ligand in order to substitute thiocyanates in the BD structure. The dipyrromethanes having 5-pentafluorophenyl and 2-thienyl substituents gave IPCE curves showing a sensitization up to 950 nm (**45**, Figure 36) [146].

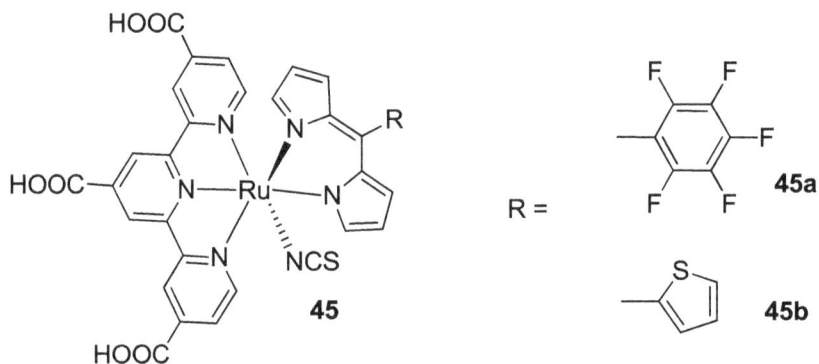

Figure 36. 2,2′-Dipyrromethane by Li *et al.* [146].

A bidentate benzimidazole was tested by Swetha *et al.* [147] as ancillary ligand in a Ru complex with tctpy, showing blueshifted absorption and a higher molecular extinction coefficient in the high energy region of the solar spectrum with respect to N749, which accounted for a better IPCE in the 400–640 nm range and a 6.07% efficiency (**46**, Figure 37; dye: 0.3 mM CH$_3$CN / *n*-butanol 1:1 with 20 mM DCA, electrolyte: 0.5 M DMPII, 0.05 M I$_2$, 0.1 M LiI CH$_3$CN / butanol 1:1).

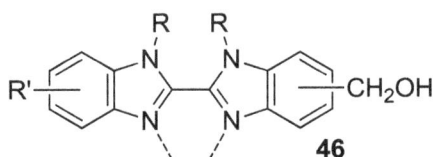

Figure 37. Benzimidazole ligand tested by Swetha *et al.* [147].

3.3. *Exchange of Metal Center*

Terpyridine complexes with other metals were reported by Bignozzi's group, who complexed osmium with tctpy, various bipyridines and pyridylquinoline [148–150]. The idea was to further broaden absorption spectra thanks to Os(II) complexes characterized by high spin-orbit coupling constant that allows the direct population of low energy, spin-forbidden, ^{3}MLCT states. No significant differences in IPCE values were found in the case of the various Os complexes showing values up to 50% at 900 nm and 70% in the visible region.

A better stability was ascribed to Os complexes with respect to the Ru case, even though these complexes showed lower light conversion.

Lapides and co-workers, in 2013 [151], tested another element of the eighth group, iron, using terpyridines as ligands in a supramolecular structure with Ru, as a multicomponent film deposed on TiO_2. An improved stability of the ruthenium dye was reported, even if these structures have not been tested on DSCs devices. More recently, Duchanois [152] reported a homoleptic iron complex bearing tridentate bis-carbene (C^N^C') ligands for sensitization of TiO_2 photoanodes (homologous to the Ru complex 43a), and compared it with a bis-tpy iron complex (47a,b, Figure 38). A considerable stabilization of ^{3}MLCT state was obtained for the cyclometalated complex, but still low performances were recorded with respect to the reference sensitizers.

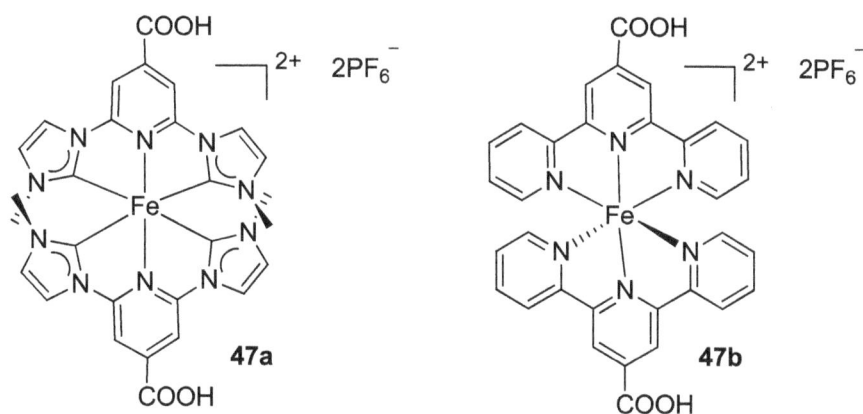

Figure 38. Iron complexes reported by Duchanois [152].

Since platinum(II) complexes usually display an intense charge-transfer absorption band in the visible region, Kwok *et al.* [153], in 2010, proposed a complex of platinum with tctpy and various alkynyl ancillary ligands, reaching up to 3.6% efficiency.

Shinpuku *et al.* [154] synthesized a series of new complexes of iridium with tpy and biphenylpyridine. Cyclometalated iridium complexes were commonly exploited in light source devices as OLEDs and showed narrower absorption spectra respect to Ruthenium ones due to more energetic MLCT transition. A shorter portion of the solar spectrum was harnessed and a lower J_{sc} was detected, nevertheless a 2.16% efficiency and long lived excited-state lifetime were reported.

The interest for d^{10} metal ions complexes such as Zn-porphyrines has grown in photonic applications, ranging from OLEDs and LECs to DSCs technologies. Bozic-Weber *et al.* [155–157] synthesized bis-tpy Zn heteroleptic complexes for TiO_2

sensitization. Terpyridines substituted with various anchoring and triphenylamino moieties extended with benzothiadiazole-diphenylamino units gave efficiencies between 0.5% and 1%. Housecroft *et al.* reviewed sensitizers made of Earth-abundant metals, concerning copper [158] and other d-block metals [159].

4. p-Type

Tpy complexes have been investigated also for the sensitization of p-type semiconductors. In p-type DSCs, the rules for sensitizers design are inverted with respect to classical n-type DSC cells. In fact, in these devices, the excited dye has to inject holes from HOMO to the conduction band of a p-semiconductor [160].

Ji *et al.*, in 2013 [161], proposed a cyclometalated (N^C^N')-(N^N'^N'') Ru[II] chromophore to sensitize NiO (**48**, Figure 39). The N^C^N' ligand was employed as anchoring moiety, while the tpy ligand was functionalized in the 4' position with a substituted naphthalenediimide (NDI) in order to withdraw electrons from the NiO surface. This dye was studied by femtosecond transient absorption spectroscopy, and results showed a slower charge recombination in the NDI-substituted complex. Perylene imides have been recently used by Sariola-Leikas *et al.* [162] as bridge groups to obtain supramolecular structures for TiO$_2$ sensitization in solid state devices.

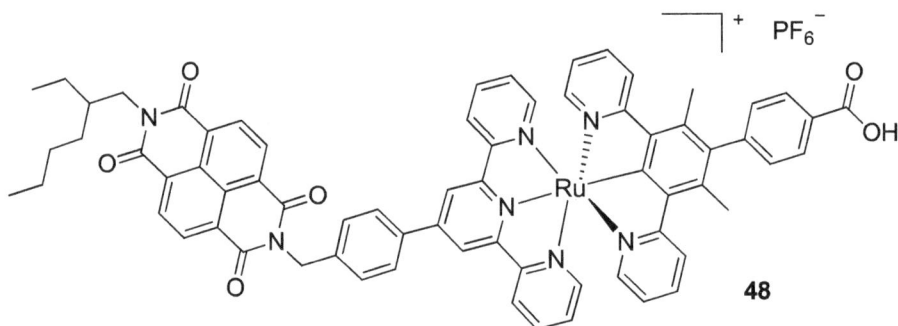

Figure 39. NDI-Tpy proposed by Ji *et al.* [161].

In 2014 both Constable [163] and Wood [164] proposed heteroleptic tpy complexes for sensitization of p-type semiconductors. The latter used a triphenyl amino moiety as anchoring donor group to increase the hole injection achieving efficiencies in pDSCs between 0.07 and 0.09 (**49**, Figure 40). Both bis-tpy and phenylbipy-tpy complexes were investigated showing better performances with iodine electrolyte with respect to the Co-based one, which was ascribed to high charge recombination with NiO.

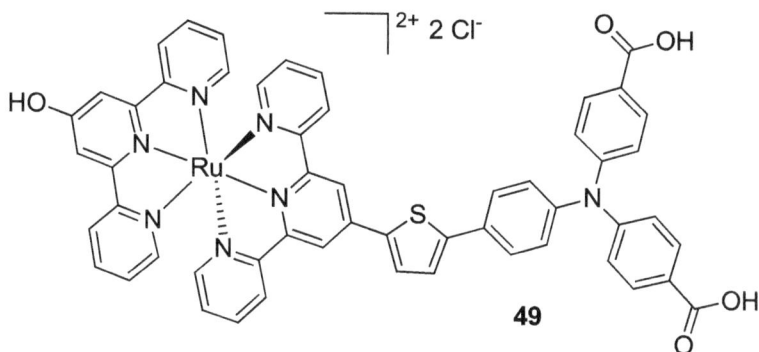

Figure 40. "K1" structure proposed by Wood *et al.* [164].

5. Co-Sensitization

The Black Dye has also been used in cocktail with other sensitizers characterized by higher molar extinction coefficient in the high energy regions of the spectrum, in order to increase the IPCE at lower wavelengths. Ogura *et al.* [165] used BD in combination with the push-pull indoline dye D131 (**50**, Figure 41), reaching a conversion efficiency of 11.0% (working electrode was made with different layers of TiO_2 mixtures with increasing amounts of polystyrene; 0.19 mM D131 and 0.56 mM BD in CH_3CN / *t*-butanol 1:1, electrolyte: 0.15 M NaI, 0.075 M I_2, 1.4 M DMPII, CH_3CN / methoxyacetonitrile 9:1). Ozawa *et al.* [166] optimized this system using 20 mM chenodeoxycholic acid achieving a 11.6% efficiency with a TiO_2 film with 45 μm thickness (0.14 mM D131 and 0.2 mM BD in 1-propanol, electrolyte: 0.05 M I_2, 0.1 M LiI, 0.6 M DMPII, 0.3 M t-bupy in CH_3CN).

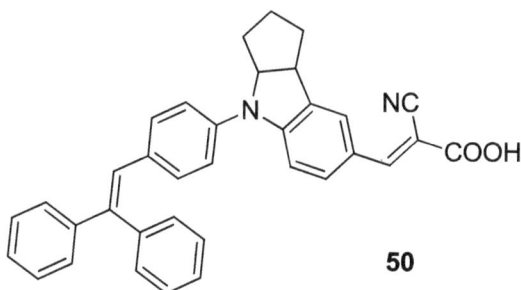

Figure 41. D131 structure used in cosensitization [165,166].

Sharma [167] proposed the cosensitization of a modified BD complex with a Zn porphyrin, with a recorded efficiency of 8.15%. Bahreman [168] synthesized a Ru complex in which a tpy was covalently bound to rhodamine B through an ethanolamine spacer, thus pursuing an energy transfer by "reverse" FRET.

6. Summary and Outlook

The literature offers multiple choices in order to tune the photoelectrochemical properties of terpyridine-based complexes such as the Black Dye, ranging from the modification of the donor and acceptor ligands to the exchange of the metal center with other cations. The increase of the molar extinction coefficient has been commonly pursued by extending the π-conjugation on the ligands. Different anchoring moieties were compared, among which COOH turned out as one of the most effective groups. Isothiocyanate was often substituted by different ancillary ligands in order to improve long-term stability and the synthetic yield of complexation; bidentate and tridentate ligands that exploit coordination through N or C atoms have been tested in order to achieve a better sensitization. Tetradentate ligands have been used in order to further enlarge the spectral absorption properties.

Few outlines can be depicted in this scenario for the design of future complexes: (1) better stability can be achieved avoiding the use of monodentate SCN ancillary ligands; (2) better performances are offered in the case of heteroleptic complexes (the homoleptic ones have an unfavourable symmetric charge distribution); (3) hydrophobic substitutions on the ligands are able to reduce the electron recombination; (4) a better coupling between the complex and semiconductor can be achieved when COOH moieties are used as attaching groups. Overall, a wise approach is requested in order to tune the energy levels far enough to reach panchromatic absorption, but not too much in order not to exceed the limit for a good regeneration rate by the electrolyte and a good electron injection driving force. Furthermore, the use of tpy complexes nowadays goes beyond the traditional role as sensitizers. Cobalt complexes have been reported as redox mediators, by exploiting the interaction of the EDOT-substituted complex with a PEDOT-covered counter-electrode (PEDOT: poly(3,4-ethylenedioxythiophene) [169]. By finely tuning the single DSC components and their interaction, a further increase of DSC performances will be possible.

Acknowledgments: The authors gratefully acknowledge financial support of the DSSCX project (PRIN 2010-2011, 20104XET32) from MIUR and Università di Torino (Ricerca Locale ex-60%, Bando 2014).

Author Contributions: DS, NB and PQ conceived and drafted the review. DS and CM screened the search results and extracted data from papers. CB and GV coordinated and supervised the project. All authors analyzed and approved the final version of the manuscript.

Conflicts of Interest: The authors declare no conflict of interest.

Abbreviations

The following abbreviations are used in this manuscript:

BD or N479	Black dye
[bmim][I]	1-butyl-3-methyl imidazolium iodide
CDCA	Chenodeoxycholic acid
DSCs	Dye sensitized Solar Cells
DCA	Deoxycholic acid
DMPII	1,2-dimethyl-3-propylimidazolium iodide
EDOT	3,4-ethylenedioxythiophene
EIS	Impedance spectroscopy
FRET	Förster Resonance Energy Transfer
GuNCS	guanidinium thiocyanate
IPCE	Incident photon to current efficiency
J_{sc}	Short circuit current
LEC	Light-emitting Electrochemical Cell
MLCT	Metal to ligand charge transfer
NDI	Naphthalenediimide
OCVD	Open Circuit Voltage Decay
OLED	Organic Light Emitting Diode
PEDOT	Poly(3,4-ethylenedioxythiophene)
qtpy	2,2':6',2":6",2'''-Quaterpyridine
SCN	Thiocyanate
TBA	Tetrabutylammonium
t-bupy	(t-butylpyridine)
tctpy	4,4',4''-Tricarboxyl-2,2':6',2''-terpyridine
TEA	Tetraethylammonium
tpy	2,2':6',2''Terpyridine
V_{oc}	Open circuit voltage

References

1. O'Regan, B.; Grätzel, M. A low-cost, high-efficiency solar cell based on dye-sensitized colloidal TiO_2 films. *Nature* **1991**, *353*, 737–740.
2. Hagfeldt, A.; Boschloo, G.; Sun, L.; Kloo, L.; Pettersson, H.; Kalyanasundaram, K. Dye-Sensitized Solar Cells. *Chem. Rev.* **2010**, *110*, 6595–6663.
3. Park, J.; Viscardi, G.; Barolo, C.; Barbero, N. Near-infrared sensitization in dye-sensitized solar cells. *Chimia* **2013**, *67*, 129–135.

4. Barbero, N.; Sauvage, F. Low Cost Electricity Production From Sunlight: Third-Generation Photovoltaics and the Dye-Sensitized Solar Cell. In *Materials for Sustainable Energy Applications: Conversion, Storage, Transmission and Consumption*; Moya, X., Munoz-Rojas, D., Eds.; CRC Press: Boca Raton, FL, USA, 2016; pp. 87–147.

5. Higashino, T.; Imahori, H. Porphyrins as excellent dyes for dye-sensitized solar cells: recent developments and insights. *Dalton Trans.* **2015**, *44*, 448–463.

6. Mishra, A.; Fischer, M.K.R.; Bäuerle, P. Metal-free organic dyes for dye-sensitized solar cells: from structure: property relationships to design rules. *Angew. Chem. Int. Ed. Engl.* **2009**, *48*, 2474–2499.

7. Abbotto, A.; Barolo, C.; Bellotto, L.; De Angelis, F.; Grätzel, M.; Manfredi, N.; Marinzi, C.; Fantacci, S.; Yum, J.-H.; Nazeeruddin, M.K. Electron-rich heteroaromatic conjugated bipyridine based ruthenium sensitizer for efficient dye-sensitized solar cells. *Chem. Commun.* **2008**, *42*, 5318–5320.

8. Saccone, D.; Galliano, S.; Barbero, N.; Viscardi, G.; Barolo, C. Polymethine dyes in hybrid photovoltaics: structure-properties relationships. *Eur. J. Org. Chem.* **2015**. in press.

9. Park, J.; Barolo, C.; Sauvage, F.; Barbero, N.; Benzi, C.; Quagliotto, P.; Coluccia, S.; Di Censo, D.; Grätzel, M.; Nazeeruddin, M.K.; *et al.* Symmetric *vs.* asymmetric squaraines as photosensitisers in mesoscopic injection solar cells: A structure–property relationship study. *Chem. Commun.* **2012**, *48*, 2782–2784.

10. Park, J.; Barbero, N.; Yoon, J.; Dell'Orto, E.; Galliano, S.; Borrelli, R.; Yum, J.-H.; Di Censo, D.; Grätzel, M.; Nazeeruddin, M.K.; *et al.* Panchromatic symmetrical squaraines: a step forward in the molecular engineering of low cost blue-greenish sensitizers for dye-sensitized solar cells. *Phys. Chem. Chem. Phys.* **2014**, *16*, 24173–24177.

11. Magistris, C.; Martiniani, S.; Barbero, N.; Park, J.; Benzi, C.; Anderson, A.; Law, C.; Barolo, C.; O'Regan, B. Near-infrared absorbing squaraine dye with extended π conjugation for dye-sensitized solar cells. *Renew. Energy* **2013**, *60*, 672–678.

12. Ono, T.; Yamaguchi, T.; Arakawa, H. Study on dye-sensitized solar cell using novel infrared dye. *Sol. Energy Mater. Sol. Cells* **2009**, *93*, 831–835.

13. Pydzińska, K.; Ziółek, M. Solar cells sensitized with near-infrared absorbing dye: Problems with sunlight conversion efficiency revealed in ultrafast laser spectroscopy studies. *Dyes Pigm.* **2015**, *122*, 272–279.

14. Nazeeruddin, M.K.; Pechy, P.; Grätzel, M. Efficient panchromatic sensitization of nanocrystalline TiO_2 films by a black dye based on a trithiocyanato-ruthenium complex. *Chem. Commun.* **1997**, *18*, 1705–1706.

15. *Scifinder, 2016*; Chemical Abstracts Service: Columbus, OH, USA, 2016. Available online: http://www.cas.org/products/scifinder (accessed on 15 January 2016).

16. Islam, A.; Sugihara, H.; Arakawa, H. Molecular design of ruthenium(II) polypyridyl photosensitizers for efficient nanocrystalline TiO_2 solar cells. *J. Photochem. Photobiol. A* **2003**, *158*, 131–138.

17. Vougioukalakis, G.C.; Philippopoulos, A.I.; Stergiopoulos, T.; Falaras, P. Contributions to the development of ruthenium-based sensitizers for dye-sensitized solar cells. *Coord. Chem. Rev.* **2011**, *255*, 2602–2621.

18. Adeloye, A.O.; Ajibade, P.A. Towards the development of functionalized polypyridine ligands for Ru(II) complexes as photosensitizers in dye-sensitized solar cells (DSSCs). *Molecules* **2014**, *19*, 12421–12460.

19. Bignozzi, C.A.; Argazzi, R.; Boaretto, R.; Busatto, E.; Carli, S.; Ronconi, F.; Caramori, S. The role of transition metal complexes in dye sensitized solar devices. *Coord. Chem. Rev.* **2013**, *257*, 1472–1492.

20. Fantacci, S.; Lobello, M.G.; De Angelis, F. Everything you always wanted to know about black dye (but were afraid to ask): A DFT/TDDFT investigation. *Chimia* **2013**, *67*, 121–128.

21. Nazeeruddin, M.K.; Péchy, P.; Renouard, T.; Zakeeruddin, S.M.; Humphry-Baker, R.; Comte, P.; Liska, P.; Cevey, L.; Costa, E.; Shklover, V.; *et al.* Engineering of efficient panchromatic sensitizers for nanocrystalline TiO_2-based solar cells. *J. Am. Chem. Soc.* **2001**, *123*, 1613–1624.

22. Wang, S.-W.; Chou, C.-C.; Hu, F.-C.; Wu, K.-L.; Chi, Y.; Clifford, J.N.; Palomares, E.; Liu, S.-H.; Chou, P.-T.; Wei, T.-C.; *et al.* Panchromatic Ru(II) sensitizers bearing single thiocyanate for high efficiency dye sensitized solar cells. *J. Mater. Chem. A* **2014**, *2*, 17618–17627.

23. Nazeeruddin, M.K.; De Angelis, F.; Fantacci, S.; Selloni, A.; Viscardi, G.; Liska, P.; Ito, S.; Takeru, B.; Grätzel, M. Combined experimental and DFT-TDDFT computational study of photoelectrochemical cell ruthenium sensitizers. *J. Am. Chem. Soc.* **2005**, *127*, 16835–16847.

24. Sauvage, F.; Decoppet, J.-D.; Zhang, M.; Zakeeruddin, S.M.; Comte, P.; Nazeeruddin, M.; Wang, P.; Grätzel, M. Effect of sensitizer adsorption temperature on the performance of dye-sensitized solar cells. *J. Am. Chem. Soc.* **2011**, *133*, 9304–9310.

25. Zakeeruddin, S.M.; Nazeeruddin, M.K.; Pechy, P.; Rotzinger, F.P.; Humphry-Baker, R.; Kalyanasundaram, K.; Grätzel, M.; Shklover, V.; Haibach, T. Molecular engineering of photosensitizers for nanocrystalline solar cells: Synthesis and characterization of Ru dyes based on phosphonated terpyridines. *Inorg. Chem.* **1997**, *36*, 5937–5946.

26. Gianotti, V.; Favaro, G.; Bonandini, L.; Palin, L.; Croce, G.; Boccaleri, E.; Artuso, E.; van Beek, W.; Barolo, C.; Milanesio, M. Rationalization of dye uptake on titania slides for dye-sensitized solar cells by a combined chemometric and structural approach. *Chem. Sus. Chem.* **2014**, *7*, 3039–3052.

27. Morgan, G.T.; Burstall, F.H. 3. Dehydrogenation of pyridine by anhydrous ferric chloride. *J. Chem. Soc.* **1932**, 20–30.

28. Heller, M.; Schubert, U.S. Syntheses of functionalized 2,2′:6′,2″-terpyridines. *Eur. J. Org. Chem.* **2003**, *2003*, 947–961.

29. Fallahpour, R.A. Synthesis of 4′-substituted-2,2′:6′, 2 -terpyridines. *Synthesis* **2003**, *35*, 155–184.

30. Cargill Thompson, A.M.W. The synthesis of 2,2′:6′,2″-terpyridine ligands — versatile building blocks for supramolecular chemistry. *Coord. Chem. Rev.* **1997**, *160*, 1–52.

31. Hayami, S.; Komatsu, Y.; Shimizu, T.; Kamihata, H.; Lee, Y.H. Spin-crossover in cobalt(II) compounds containing terpyridine and its derivatives. *Coord. Chem. Rev.* **2011**, *255*, 1981–1990.

32. Arrigo, A.; Santoro, A.; Puntoriero, F.; Lainé, P.P.; Campagna, S. Photoinduced electron transfer in donor–bridge–acceptor assemblies: The case of Os(II)-bis(terpyridine)-(bi)pyridinium dyads. *Coord. Chem. Rev.* **2015**, *304-305*, 109–116.

33. Sauvage, J.P.; Collin, J.P.; Chambron, J.C.; Guillerez, S.; Coudret, C.; Balzani, V.; Barigelletti, F.; De Cola, L.; Flamigni, L. Ruthenium(II) and Osmium(II) Bis(terpyridine) Complexes in Covalently-Linked Multicomponent Systems: Synthesis, Electrochemical Behavior, Absorption Spectra, and Photochemical and Photophysical Properties. *Chem. Rev.* **1994**, *94*, 993–1019.

34. Flamigni, L.; Collin, J.P.; Sauvage, J.P. Iridium terpyridine complexes as functional assembling units in arrays for the conversion of light energy. *Acc. Chem. Res.* **2008**, *41*, 857–871.

35. Baranoff, E.; Collin, J.P.; Flamigni, L.; Sauvage, J.P. From ruthenium(II) to iridium(III): 15 years of triads based on bis-terpyridine complexes. *Chem. Soc. Rev.* **2004**, *33*, 147–155.

36. Eryazici, I.; Moorefield, C.N.; Newkome, G.R. Square-planar Pd(II), Pt(II), and Au(III) terpyridine complexes: their syntheses, physical properties, supramolecular constructs, and biomedical activities. *Chem. Rev.* **2008**, *108*, 1834–1895.

37. Hofmeier, H.; Schubert, U.S. Recent developments in the supramolecular chemistry of terpyridine-metal complexes. *Chem. Soc. Rev.* **2004**, *33*, 373–399.

38. Schubert, U.S.; Hofmeier, H.; Newkome, G.R. *Modern Terpyridine Chemistry*; John Wiley & Sons, Inc.: Hoboken, NJ, USA, 2006.

39. Gao, Y.; Rajwar, D.; Grimsdale, A.C. Self-Assembly of Conjugated Units Using Metal-Terpyridine Coordination. *Macromol. Rapid Commun.* **2014**, *35*, 1727–1740.

40. Constable, E.C. 2,2′:6′,2″-Terpyridines: From chemical obscurity to common supramolecular motifs. *Chem. Soc. Rev.* **2007**, *36*, 246–253.

41. Sakamoto, R.; Wu, K.-H.; Matsuoka, R.; Maeda, H.; Nishihara, H. π-Conjugated bis(terpyridine)metal complex molecular wires. *Chem. Soc. Rev.* **2015**, *44*, 7698–7714.

42. Andres, P.R.; Schubert, U.S. New Functional Polymers and Materials Based on 2,2′:6′,2″-Terpyridine Metal Complexes. *Adv. Mater.* **2004**, *16*, 1043–1068.

43. Winter, A.; Hoeppener, S.; Newkome, G.R.; Schubert, U.S. Terpyridine-functionalized surfaces: redox-active, switchable, and electroactive nanoarchitectures. *Adv. Mater.* **2011**, *23*, 3484–3498.

44. Breivogel, A.; Kreitner, C.; Heinze, K. Redox and Photochemistry of Bis(terpyridine)ruthenium(II) Amino Acids and Their Amide Conjugates - from Understanding to Applications. *Eur. J. Inorg. Chem.* **2014**, *2014*, 5468–5490.

45. Winter, A.; Gottschaldt, M.; Newkome, G.R.; Schubert, U.S. Terpyridines and their Complexes with First Row Transition Metal Ions: Cytotoxicity, Nuclease Activity and Self-Assembly of Biomacromolecules. *Curr. Top. Med. Chem.* **2012**, *12*, 158–175.

46. Winter, A.; Hager, M.D.; Newkome, G.R.; Schubert, U.S. The marriage of terpyridines and inorganic nanoparticles: synthetic aspects, characterization techniques, and potential applications. *Adv. Mater.* **2011**, *23*, 5728–5748.

47. Wild, A.; Winter, A.; Schlütter, F.; Schubert, U.S. Advances in the field of π-conjugated 2,2′:6′,2″-terpyridines. *Chem. Soc. Rev.* **2011**, *40*, 1459–1511.

48. Chelucci, G.; Thummel, R.P. Chiral 2,2′-Bipyridines, 1,10-Phenanthrolines, and 2,2′:6′,2″-Terpyridines: Syntheses and Applications in Asymmetric Homogeneous Catalysis. *Chem. Rev.* **2002**, *102*, 3129–3170.

49. Winter, A.; Newkome, G.R.; Schubert, U.S. Catalytic Applications of Terpyridines and their Transition Metal Complexes. *ChemCatChem* **2011**, *3*, 1384–1406.

50. Kröhnke, F. The Specific Synthesis of Pyridines and Oligopyridines. *Synthesis* **1976**, *1*, 1–24.

51. Cave, G.W.V.; Raston, C.L. Efficient synthesis of pyridines via a sequential solventless aldol condensation and Michael addition. *J. Chem. Soc. Perkin Trans. 1* **2001**, 3258–3264.

52. Suzuki, A. Cross-coupling reactions of organoboranes: an easy way to construct C-C bonds (Nobel Lecture). *Angew. Chem. Int. Ed. Engl.* **2011**, *50*, 6722–6737.

53. Cordovilla, C.; Bartolomé, C.; Martínez-Ilarduya, J.M.; Espinet, P. The Stille Reaction, 38 Years Later. *ACS Catal.* **2015**, *5*, 3040–3053.

54. Nazeeruddin, M.K.; Pechy, P.; Renouard, T.; Zakeeruddin, S.M.; Humphry-Baker, R.; Comte, P.; Liska, P.; Cevey, L.; Costa, E.; Shklover, V.; Spiccia, L.; Deacon, G.B.; Bignozzi, C.A.; Grätzel, M. Engineering of Efficient Panchromatic Sensitizers for Nanocrystalline TiO$_2$-Based Solar Cells. *J. Am. Chem. Soc* **2001**, *123*, 1613–1624.

55. Dehaudt, J.; Husson, J.; Guyard, L. A more efficient synthesis of 4,4′,4″-tricarboxy-2,2′:6′,2″-terpyridine. *Green Chem.* **2011**, *13*, 3337–3340.

56. Husson, J.; Knorr, M. 2,2′:6′,2″-Terpyridines Functionalized with Thienyl Substituents: Synthesis and Applications. *J. Heterocycl. Chem.* **2012**, *49*, 453–478.

57. Kisserwan, H.; Kamar, A.; Shoker, T.; Ghaddar, T.H. Photophysical properties of new cyclometalated ruthenium complexes and their use in dye sensitized solar cells. *Dalton Trans.* **2012**, *41*, 10643–10651.

58. Raboin, J.-C.; Kirsch, G.; Beley, M. On the way to unsymmetrical terpyridines carrying carboxylic acids. *J. Heterocycl. Chem.* **2000**, *37*, 1077–1080.

59. Husson, J.; Beley, M.; Kirsch, G. A novel pathway for the synthesis of a carboxylic acid-functionalised Ru(II) terpyridine complex. *Tetrahedron Lett.* **2003**, *44*, 1767–1770.

60. Husson, J.; Dehaudt, J.; Guyard, L. Preparation of carboxylate derivatives of terpyridine via the furan pathway. *Nat. Protoc.* **2014**, *9*, 21–26.

61. Hobert, S.E.; Carney, J.T.; Cummings, S.D. Synthesis and luminescence properties of platinum(II) complexes of 4′-chloro-2,2′:6′,2″-terpyridine and 4,4′,4″-trichloro-2,2′:6′,2″-terpyridine. *Inorganica Chim. Acta* **2001**, *318*, 89–96.

62. Duncan, T.V.; Ishizuka, T.; Therien, M.J. Molecular engineering of intensely near-infrared absorbing excited states in highly conjugated oligo(porphinato)zinc-(polypyridyl)metal(II) supermolecules. *J. Am. Chem. Soc.* **2007**, *129*, 9691–9703.

63. Potts, K.T.; Konwar, D. Synthesis of 4'-Vinyl-2,2':6',2''-terpyridine. *J. Org. Chem.* **1991**, *56*, 4815–4816.

64. Husson, J.; Knorr, M. Syntheses and applications of furanyl-functionalised 2,2':6',2''-Terpyridines. *Beilstein J. Org. Chem.* **2012**, *8*, 379–389.

65. Woodward, C.P.; Coghlan, C.J.; Rüther, T.; Jones, T.W.; Hebting, Y.; Cordiner, R.L.; Dawson, R.E.; Robinson, D.E.J.E.; Wilson, G.J. Oligopyridine ligands possessing multiple or mixed anchoring functionality for dye-sensitized solar cells. *Tetrahedron* **2015**, *71*, 5238–5247.

66. Dick, G.R.; Woerly, E.M.; Burke, M.D. A general solution for the 2-pyridyl problem. *Angew. Chem. Int. Ed. Engl.* **2012**, *51*, 2667–2672.

67. Coluccini, C.; Manfredi, N.; Salamone, M.M.; Ruffo, R.; Lobello, M.G.; De Angelis, F.; Abbotto, A. Quaterpyridine ligands for panchromatic Ru(II) dye sensitizers. *J. Org. Chem.* **2012**, *77*, 7945–7956.

68. Barolo, C.; Yum, J.-H.; Artuso, E.; Barbero, N.; Di Censo, D.; Lobello, M.G.; Fantacci, S.; De Angelis, F.; Grätzel, M.; Nazeeruddin, M.K.; Viscardi, G. A simple synthetic route to obtain pure trans-ruthenium(II) complexes for dye-sensitized solar cell applications. *ChemSusChem* **2013**, *6*, 2170–2180.

69. Waser, M.; Siebenhaar, C.; Zampese, J.; Grundler, G.; Constable, E.; Height, M.; Pieles, U. Novel grafting procedure of ruthenium 2,2':6',2''-terpyridine complexes with phosphonate ligands to titania for water splitting applications. *Chimia* **2010**, *64*, 328–329.

70. Anthonysamy, A.; Balasubramanian, S.; Muthuraaman, B.; Maruthamuthu, P. 4'-functionalized 2,2':6',2'' terpyridine ruthenium (II) complex: a nanocrystalline TiO_2 based solar cell sensitizer. *Nanotechnology* **2007**, *18*, 095701/1–095701/5.

71. Wang, Z.-S.; Huang, C.-H.; Huang, Y.-Y.; Zhang, B.-W.; Xie, P.-H.; Hou, Y.-J.; Ibrahim, K.; Qian, H.-J.; Liu, F.-Q. Photoelectric behavior of nanocrystalline TiO_2 electrode with a novel terpyridyl ruthenium complex. *Sol. Energy Mater. Sol. Cells* **2002**, *71*, 261–271.

72. Funaki, T.; Yanagida, M.; Onozawa-Komatsuzaki, N.; Kawanishi, Y.; Kasuga, K.; Sugihara, H. Ruthenium (II) complexes with π expanded ligand having phenylene–ethynylene moiety as sensitizers for dye-sensitized solar cells. *Sol. Energy Mater. Sol. Cells* **2009**, *93*, 729–732.

73. McNamara, W.R.; Snoeberger III, R.C.; Li, G.; Richter, C.; Allen, L.J.; Milot, R.L.; Schmuttenmaer, C.A.; Crabtree, R.H.; Brudvig, G.W.; Batista, V.S. Hydroxamate anchors for water-stable attachment to TiO_2 nanoparticles. *Energy Environ. Sci.* **2009**, *2*, 1173–1175.

74. Vougioukalakis, G.C.; Stergiopoulos, T.; Kantonis, G.; Kontos, A.G.; Papadopoulos, K.; Stublla, A.; Potvin, P.G.; Falaras, P. Terpyridine- and 2,6-dipyrazinylpyridine-coordinated ruthenium(II) complexes: Synthesis, characterization and application in TiO_2-based dye-sensitized solar cells. *J. Photochem. Photobiol., A* **2010**, *214*, 22–32.

75. Manríquez, J.; Hwang, S.-H.; Cho, T.J.; Moorefield, C.N.; Newkome, G.R.; Godínez, L.A. Sensitized Solar Cells based on Hexagonal Dyes of Terpyridine-Ruthenium(II): Effect of the Electropolymerization of Dyes during their Performance in Solar Cells. *ECS Trans.* **2006**, *3*, 1–5.

76. Kanniyambatti Lourdusamy, V.J.; Anthonysamy, A.; Easwaramoorthi, R.; Shinde, D.V.; Ganapathy, V.; Karthikeyan, S.; Lee, J.; Park, T.; Rhee, S.-W.; Kim, K.S.; Kim, J.K. Cyanoacetic acid tethered thiophene for well-matched LUMO level in Ru(II)-terpyridine dye sensitized solar cells. *Dyes Pigm.* **2016**, *126*, 270–278.

77. Numata, Y.; Singh, S.P.; Islam, A.; Iwamura, M.; Imai, A.; Nozaki, K.; Han, L. Enhanced Light-Harvesting Capability of a Panchromatic Ru(II) Sensitizer Based on π-Extended Terpyridine with a 4-Methylstylryl Group for Dye-Sensitized Solar Cells. *Adv. Funct. Mater.* **2013**, *23*, 1817–1823.

78. Yang, S.-H.; Wu, K.-L.; Chi, Y.; Cheng, Y.-M.; Chou, P.-T. Tris(thiocyanate) ruthenium(II) sensitizers with functionalized dicarboxyterpyridine for dye-sensitized solar cells. *Angew. Chem. Int. Ed. Engl.* **2011**, *50*, 8270–8274.

79. Kimura, M.; Masuo, J.; Tohata, Y.; Obuchi, K.; Masaki, N.; Murakami, T.N.; Koumura, N.; Hara, K.; Fukui, A.; Yamanaka, R.; Mori, S. Improvement of TiO$_2$/dye/electrolyte interface conditions by positional change of alkyl chains in modified panchromatic Ru complex dyes. *Chem. - Eur. J.* **2013**, *19*, 1028–1034.

80. Dehaudt, J.; Husson, J.; Guyard, L.; Oswald, F.; Martineau, D. A simple access to "Black-Dye" analogs with good efficiencies in dye-sensitized solar cells. *Renew. Energy* **2014**, *66*, 588–595.

81. Koyyada, G.; Botla, V.; Thogiti, S.; Wu, G.; Li, J.; Fang, X.; Kong, F.; Dai, S.; Surukonti, N.; Kotamarthi, B.; Malapaka, C. Novel 4′-functionalized 4,4″-dicarboxyterpyridine ligands for ruthenium complexes: near-IR sensitization in dye sensitized solar cells. *Dalton Trans.* **2014**, *43*, 14992–15003.

82. Ozawa, H.; Fukushima, K.; Sugiura, T.; Urayama, A.; Arakawa, H. Ruthenium sensitizers having an ortho-dicarboxyl group as an anchoring unit for dye-sensitized solar cells: synthesis, photo- and electrochemical properties, and adsorption behavior to the TiO$_2$ surface. *Dalton Trans.* **2014**, *43*, 13208–13218.

83. Ozawa, H.; Sugiura, T.; Shimizu, R.; Arakawa, H. Novel ruthenium sensitizers having different numbers of carboxyl groups for dye-sensitized solar cells: effects of the adsorption manner at the TiO$_2$ surface on the solar cell performance. *Inorg. Chem.* **2014**, *53*, 9375–9384.

84. Ozawa, H.; Yamamoto, Y.; Kawaguchi, H.; Shimizu, R.; Arakawa, H. Ruthenium sensitizers with a hexylthiophene-modified terpyridine ligand for dye-sensitized solar cells: synthesis, photo- and electrochemical properties, and adsorption behavior to the TiO$_2$ surface. *ACS Appl. Mater. Interfaces* **2015**, *7*, 3152–3161.

85. Ozawa, H.; Fukushima, K.; Urayama, A.; Arakawa, H. Efficient ruthenium sensitizer with an extended π-conjugated terpyridine ligand for dye-sensitized solar cells. *Inorg. Chem.* **2015**, *54*, 8887–8889.

86. Ozawa, H.; Yamamoto, Y.; Fukushima, K.; Yamashita, S.; Arakawa, H. Synthesis and Characterization of a Novel Ruthenium Sensitizer with a Hexylthiophene-functionalized Terpyridine Ligand for Dye-sensitized Solar Cells. *Chem. Lett.* **2013**, *42*, 897–899.

87. Ozawa, H.; Kuroda, T.; Harada, S.; Arakawa, H. Efficient Ruthenium Sensitizer with a Terpyridine Ligand Having a Hexylthiophene Unit for Dye-Sensitized Solar Cells: Effects of the Substituent Position on the Solar Cell Performance. *Eur. J. Inorg. Chem.* **2014**, *2014*, 4734–4739.

88. Nazeeruddin, M.K.; Zakeeruddin, S.M.; Humphry-Baker, R.; Gorelsky, S.I.; Lever, A.B.P.; Grätzel, M. Synthesis, spectroscopic and a ZINDO study of cis- and trans-(X_2)bis(4,4'-dicarboxylic acid-2,2'-bipyridine)ruthenium(II) complexes $(X=Cl^-, H_2O, NCS^-)$. *Coord. Chem. Rev.* **2000**, *208*, 213–225.

89. Renouard, T.; Grätzel, M. Functionalized tetradentate ligands for Ru-sensitized solar cells. *Tetrahedron* **2001**, *57*, 8145–8150.

90. Renouard, T.; Fallahpour, R.-A.; Nazeeruddin, M.K.; Humphry-Baker, R.; Gorelsky, S.I.; Lever, A.B.P.; Grätzel, M. Novel Ruthenium Sensitizers Containing Functionalized Hybrid Tetradentate Ligands: Synthesis, Characterization, and INDO/S Analysis. *Inorg. Chem.* **2002**, *41*, 367–378.

91. Barolo, C.; Nazeeruddin, M.K.; Fantacci, S.; Di Censo, D.; Comte, P.; Liska, P.; Viscardi, G.; Quagliotto, P.; De Angelis, F.; Ito, S.; Grätzel, M. Synthesis, characterization, and DFT-TDDFT computational study of a ruthenium complex containing a functionalized tetradentate ligand. *Inorg. Chem.* **2006**, *45*, 4642–4653.

92. Abbotto, A.; Sauvage, F.; Barolo, C.; De Angelis, F.; Fantacci, S.; Grätzel, M.; Manfredi, N.; Marinzi, C.; Nazeeruddin, M.K. Panchromatic ruthenium sensitizer based on electron-rich heteroarylvinylene π-conjugated quaterpyridine for dye-sensitized solar cells. *Dalton Trans.* **2011**, *40*, 234–242.

93. *Dye-sensitized Solar Cells*; Kalyanasundaram, K., Ed.; CRC Press: Boca Raton, Florida, US, 2010.

94. Nazeeruddin, M.K.; Grätzel, M. Separation of linkage isomers of trithiocyanato (4,4',4"-tricarboxy-2,2':6,2"-terpyridine)ruthenium(II) by pH-titration method and their application in nanocrystalline TiO_2-based solar cells. *J. Photochem. Photobiol., A* **2001**, *145*, 79–86.

95. Pal, A.K.; Hanan, G.S. Design, synthesis and excited-state properties of mononuclear Ru(II) complexes of tridentate heterocyclic ligands. *Chem. Soc. Rev.* **2014**, *43*, 6184–6197.

96. Medlycott, E.A.; Hanan, G.S. Designing tridentate ligands for ruthenium(II) complexes with prolonged room temperature luminescence lifetimes. *Chem. Soc. Rev.* **2005**, *34*, 133–142.

97. Hammarström, L.; Johansson, O. Expanded bite angles in tridentate ligands. Improving the photophysical properties in bistridentate Ru (II) polypyridine complexes. *Coord. Chem. Rev.* **2010**, *254*, 2546–2559.

98. Chandrasekharam, M.; Rajkumar, G.; Rao, C.S.; Suresh, T.; Soujanya, Y.; Reddy, P.Y. High molar extinction coefficient Ru(II)-mixed ligand polypyridyl complexes for dye sensitized solar cell application. *Adv. Optoelectron.* **2011**, *2011*, 1–12.

99. Giribabu, L.; Singh, V.K.; Srinivasu, M.; Kumar, C.V.; Reddy, V.G.; Soujnya, Y.; Reddy, P.Y. Synthesis and photoelectrochemical characterization of a high molar extinction coefficient heteroleptic ruthenium(II) complex. *J. Chem. Sci.* **2011**, *123*, 371–378.

100. Koyyada, G.; Pavan Kumar, CH.; Salvatori, P.; Marotta, G.; Lobello, M.G.; Bizzarri, O.; De Angelis, F.; Malapaka, C. New terpyridine-based ruthenium complexes for dye sensitized solar cells applications. *Inorganica Chim. Acta* **2016**, *442*, 158–166.

101. Pavan Kumar, C.H.; Anusha, V.; Narayanaswamy, K.; Bhanuprakash, K.; Islam, A.; Han, L.; Singh, S.P.; Chandrasekharam, M. New ruthenium complexes (Ru[3+2+1]) bearing π-extended 4-methylstyryl terpyridine and unsymmetrical bipyridine ligands for DSSC applications. *Inorganica Chim. Acta* **2015**, *435*, 46–52.

102. Giribabu, L.; Bessho, T.; Srinivasu, M.; Vijaykumar, C.; Soujanya, Y.; Reddy, V.G.; Reddy, P.Y.; Yum, J.-H.; Grätzel, M.; Nazeeruddin, M.K. A new family of heteroleptic ruthenium(II) polypyridyl complexes for sensitization of nanocrystalline TiO_2 films. *Dalton Trans.* **2011**, *40*, 4497–4504.

103. Mosurkal, R.; Kim, Y.; Kumar, J.; Li, L.; Walker, J.; Samuelson, L.A. Mono- and Dinuclear Ruthenium Complexes for Nanocrystalline TiO_2 Based Dye-Sensitized Photovoltaics. *J. Macromol. Sci. Part A* **2003**, *40*, 1317–1325.

104. Erten-Ela, S.; Sogut, S.; Ocakoglu, K. Synthesis of novel ruthenium II phenanthroline complex and its application to TiO_2 and ZnO nanoparticles on the electrode of dye sensitized solar cells. *Mater. Sci. Semicond. Process.* **2014**, *23*, 159–166.

105. Mongal, B.N.; Pal, A.; Mandal, T.K.; Datta, J.; Naskar, S. Synthesis, characterisation, electrochemical study and photovoltaic measurements of a new terpyridine and pyridine-quinoline based mixed chelate ruthenium dye. *Polyhedron* **2015**, *102*, 615–626.

106. Stergiopoulos, T.; Arabatzis, I.M.; Kalbac, M.; Lukes, I.; Falaras, P. Incorporation of innovative compounds in nanostructured photoelectrochemical cells. *J. Mater. Process. Technol.* **2005**, *161*, 107–112.

107. Houarner, C.; Blart, E.; Buvat, P.; Odobel, F. Ruthenium bis-terpyridine complexes connected to an oligothiophene unit for dry dye-sensitised solar cells. *Photochem. Photobiol. Sci.* **2005**, *4*, 200–204.

108. Houarner-Rassin, C.; Blart, E.; Buvat, P.; Odobel, F. Improved efficiency of a thiophene linked ruthenium polypyridine complex for dry dye-sensitized solar cells. *J. Photochem. Photobiol. A* **2007**, *186*, 135–142.

109. Houarner-Rassin, C.; Chaignon, F.; She, C.; Stockwell, D.; Blart, E.; Buvat, P.; Lian, T.; Odobel, F. Synthesis and photoelectrochemical properties of ruthenium bisterpyridine sensitizers functionalized with a thienyl phosphonic acid moiety. *J. Photochem. Photobiol. A* **2007**, *192*, 56–65.

110. Duprez, V.; Biancardo, M.; Krebs, F.C. Characterisation and application of new carboxylic acid-functionalised ruthenium complexes as dye-sensitisers for solar cells. *Sol. Energy Mater. Sol. Cells* **2007**, *91*, 230–237.

111. Duprez, V.; Krebs, F.C. New carboxy-functionalized terpyridines as precursors for zwitterionic ruthenium complexes for polymer-based solar cells. *Tetrahedron Lett.* **2006**, *47*, 3785–3789.

112. Duprez, V.; Biancardo, M.; Spanggaard, H.; Krebs, F.C. Synthesis of Conjugated Polymers Containing Terpyridine–Ruthenium Complexes: Photovoltaic Applications. *Macromolecules* **2005**, *38*, 10436–10448.

113. Krebs, F.C.; Biancardo, M. Dye sensitized photovoltaic cells: Attaching conjugated polymers to zwitterionic ruthenium dyes. *Sol. Energy Mater. Sol. Cells* **2006**, *90*, 142–165.

114. Chan, H.T.; Mak, C.S.K.; Djurišić, A.B.; Chan, W.K. Synthesis of Ruthenium Complex Containing Conjugated Polymers and Their Applications in Dye-Sensitized Solar Cells. *Macromol. Chem. Phys.* **2011**, *212*, 774–784.

115. Caramori, S.; Husson, J.; Beley, M.; Bignozzi, C. A.; Argazzi, R.; Gros, P.C. Combination of cobalt and iron polypyridine complexes for improving the charge separation and collection in Ru(terpyridine)(2)-sensitised solar cells. *Chem. - Eur. J.* **2010**, *16*, 2611–2618.

116. Funaki, T.; Funakoshi, H.; Kitao, O.; Onozawa-Komatsuzaki, N.; Kasuga, K.; Sayama, K.; Sugihara, H. Cyclometalated ruthenium(II) complexes as near-IR sensitizers for high efficiency dye-sensitized solar cells. *Angew. Chem. Int. Ed. Engl.* **2012**, *51*, 7528–7531.

117. Kusama, H.; Funaki, T.; Sayama, K. Theoretical study of cyclometalated Ru(II) dyes: Implications on the open-circuit voltage of dye-sensitized solar cells. *J. Photochem. Photobiol., A* **2013**, *272*, 80–89.

118. Funaki, T.; Yanagida, M.; Onozawa-Komatsuzaki, N.; Kasuga, K.; Kawanishi, Y.; Kurashige, M.; Sayama, K.; Sugihara, H. Synthesis of a new class of cyclometallated ruthenium(II) complexes and their application in dye-sensitized solar cells. *Inorg. Chem. Commun.* **2009**, *12*, 842–845.

119. Islam, A.; Singh, S.P.; Han, L. Synthesis and application of new ruthenium complexes containing β-diketonato ligands as sensitizers for nanocrystalline TiO_2 solar cells. *Int. J. Photoenergy* **2011**, *2011*, 204639.

120. Han, L.; Islam, A. High efficient dye-sensitized solar cells. *MRS Online Proc. Libr.* **2011**, *1327*. Symposium G - Complex Oxide Materials for Emerging Energy Technologies.

121. Islam, A.; Singh, S.P.; Han, L. Thiocyanate-free, panchromatic ruthenium (II) terpyridine sensitizer having a tridentate diethylenetriamine ligand for Near-IR sensitization of nanocrystaline TiO_2. *Funct. Mater. Lett.* **2011**, *04*, 21–24.

122. Islam, A.; Singh, S.P.; Yanagida, M.; Karim, M.R.; Han, L. Amphiphilic ruthenium(II) terpyridine sensitizers with long alkyl chain substituted beta-diketonato ligands: An efficient coadsorbent-free dye-sensitized solar cells. *Int. J. Photoenergy*, 2011.

123. Islam, A.; Chowdhury, F.A.; Chiba, Y.; Komiya, R.; Fuke, N.; Ikeda, N.; Nozaki, K.; Han, L. Synthesis and Characterization of New Efficient Tricarboxyterpyridyl (β-diketonato) Ruthenium(II) Sensitizers and Their Applications in Dye-Sensitized Solar Cells. *Chem. Mater.* **2006**, *18*, 5178–5185.

124. Islam, A.; Chowdhury, F.A.; Chiba, Y.; Komiya, R.; Fuke, N.; Ikeda, N.; Han, L. Ruthenium(II) tricarboxyterpyridyl complex with a fluorine-substituted β-diketonato ligand for highly efficient dye-sensitized solar cells. *Chem. Lett.* **2005**, *34*, 344–345.

125. Islam, A.; Sugihara, H.; Yanagida, M.; Hara, K.; Fujihashi, G.; Tachibana, Y.; Katoh, R.; Murata, S.; Arakawa, H. Efficient panchromatic sensitization of nanocrystalline TiO_2 films by beta-diketonato ruthenium polypyridyl complexes. *New J. Chem.* **2002**, *26*, 966–968.

126. Jiang, X.; Marinado, T.; Gabrielsson, E.; Hagberg, D.P.; Sun, L.; Hagfeldt, A. Structural Modification of Organic Dyes for Efficient Coadsorbent-Free Dye-Sensitized Solar Cells. *J. Phys. Chem. C* **2010**, *114*, 2799–2805.

127. Chen, B.-S.; Chen, K.; Hong, Y.-H.; Liu, W.-H.; Li, T.-H.; Lai, C.-H.; Chou, P.-T.; Chi, Y.; Lee, G.-H. Neutral, panchromatic Ru(II) terpyridine sensitizers bearing pyridine pyrazolate chelates with superior DSSC performance. *Chem. Commun.* **2009**, 5844–5846.

128. Chou, C.-C.; Wu, K.-L.; Chi, Y.; Hu, W.-P.; Yu, S.J.; Lee, G.-H.; Lin, C.-L.; Chou, P.-T. Ruthenium(II) sensitizers with heteroleptic tridentate chelates for dye-sensitized solar cells. *Angew. Chem. Int. Ed. Engl.* **2011**, *50*, 2054–2058.

129. Wu, K.-L.; Li, C.-H.; Chi, Y.; Clifford, J.N.; Cabau, L.; Palomares, E.; Cheng, Y.-M.; Pan, H.-A.; Chou, P.-T. Dye molecular structure device open-circuit voltage correlation in Ru(II) sensitizers with heteroleptic tridentate chelates for dye-sensitized solar cells. *J. Am. Chem. Soc* **2012**, *134*, 7488–7496.

130. Chou, C.-C.; Hu, F.-C.; Yeh, H.-H.; Wu, H.-P.; Chi, Y.; Clifford, J. N.; Palomares, E.; Liu, S.-H.; Chou, P.-T.; Lee, G.-H. Highly efficient dye-sensitized solar cells based on panchromatic ruthenium sensitizers with quinolinylbipyridine anchors. *Angew. Chem. Int. Ed. Engl.* **2014**, *53*, 178–183.

131. Chang, T.-K.; Li, H.; Chen, K.-T.; Tsai, Y.-C.; Chi, Y.; Hsiao, T.-Y.; Kai, J.-J. Substituent effect of Ru(II)-based sensitizers bearing a terpyridine anchor and a pyridyl azolate ancillary for dye sensitized solar cells. *J. Mater. Chem. A* **2015**, *3*, 18422–18431.

132. Wadman, S.H.; Kroon, J.M.; Bakker, K.; Lutz, M.; Spek, A.L.; van Klink, G.P.M.; van Koten, G. Cyclometalated ruthenium complexes for sensitizing nanocrystalline TiO$_2$ solar cells. *Chem. Commun.* **2007**, 1907–1909.

133. Wadman, S.H.; Kroon, J.M.; Bakker, K.; Havenith, R.W.A.; van Klink, G.P.M.; van Koten, G. Cyclometalated Organoruthenium Complexes for Application in Dye-Sensitized Solar Cells. *Organometallics* **2010**, *29*, 1569–1579.

134. Wadman, S.H.; van Leeuwen, Y.M.; Havenith, R.W.A.; van Klink, G.P.M.; van Koten, G. A Redox Asymmetric, Cyclometalated Ruthenium Dimer: Toward Upconversion Dyes in Dye-Sensitized TiO$_2$ Solar Cells. *Organometallics* **2010**, *29*, 5635–5645.

135. Kisserwan, H.; Ghaddar, T.H. Enhancement of photocurrent in dye sensitized solar cells incorporating a cyclometalated ruthenium complex with cuprous iodide as an electrolyte additive. *Dalton Trans.* **2011**, *40*, 3877–3884.

136. Robson, K.C.D.; Koivisto, B.D.; Yella, A.; Sporinova, B.; Nazeeruddin, M.K.; Baumgartner, T.; Grätzel, M.; Berlinguette, C.P. Design and development of functionalized cyclometalated ruthenium chromophores for light-harvesting applications. *Inorg. Chem.* **2011**, *50*, 5494–5508.

137. Al-mutlaq, F.A.; Potvin, P.G.; Philippopoulos, A.I.; Falaras, P. Catechol-Bearing Dipyrazinylpyridine Complexes of Ruthenium(II). *Eur. J. Inorg. Chem.* **2007**, *2007*, 2121–2128.

138. Sepehrifard, A.; Chen, S.; Stublla, A.; Potvin, P.G.; Morin, S. Effects of ligand LUMO levels, anchoring groups and spacers in Ru(II)-based terpyridine and dipyrazinylpyridine complexes on adsorption and photoconversion efficiency in DSSCs. *Electrochim. Acta* **2013**, *87*, 236–244.

139. Sepehrifard, A.; Stublla, A.; Haftchenary, S.; Chen, S.; Potvin, P.G.; Morin, S. Effects of carboxyl and ester anchoring groups on solar conversion efficiencies of TiO_2 dye-sensitized solar cells. *J. New. Mat. Electrochem. Syst.* **2008**, *11*, 281–285.

140. Schulze, B.; Brown, D.G.; Robson, K.C.D.; Friebe, C.; Jäger, M.; Birckner, E.; Berlinguette, C.P.; Schubert, U.S. Cyclometalated ruthenium(II) complexes featuring tridentate click-derived ligands for dye-sensitized solar cell applications. *Chem. - Eur. J.* **2013**, *19*, 14171–14180.

141. Sinn, S.; Schulze, B.; Friebe, C.; Brown, D.G.; Jäger, M.; Kübel, J.; Dietzek, B.; Berlinguette, C.P.; Schubert, U.S. A heteroleptic bis(tridentate) ruthenium(II) platform featuring an anionic 1,2,3-triazolate-based ligand for application in the dye-sensitized solar cell. *Inorg. Chem.* **2014**, *53*, 1637–1645.

142. Park, H.-J.; Kim, K.H.; Choi, S.Y.; Kim, H.-M.; Lee, W.I.; Kang, Y.K.; Chung, Y.K. Unsymmetric Ru(II) Complexes with N-Heterocyclic Carbene and/or Terpyridine Ligands: Synthesis, Characterization, Ground- and Excited-State Electronic Structures and Their Application for DSSC Sensitizers. *Inorg. Chem.* **2010**, *49*, 7340–7352.

143. Bonacin, J.A.; Toma, S.H.; Freitas, J.N.; Nogueira, A.F.; Toma, H.E. On the behavior of the carboxyphenylterpyridine(8-quinolinolate) thiocyanatoruthenium(II) complex as a new black dye in TiO_2 solar cells modified with carboxymethyl-beta-cyclodextrin. *Inorg. Chem. Commun.* **2013**, *36*, 35–38.

144. Kinoshita, T.; Dy, J.T.; Uchida, S.; Kubo, T.; Segawa, H. Wideband dye-sensitized solar cells employing a phosphine-coordinated ruthenium sensitizer. *Nat. Photonics* **2013**, *7*, 535–539.

145. Kinoshita, T.; Nonomura, K.; Joong Jeon, N.; Giordano, F.; Abate, A.; Uchida, S.; Kubo, T.; Seok, S.I.; Nazeeruddin, M.K.; Hagfeldt, A.; *et al.* Spectral splitting photovoltaics using perovskite and wideband dye-sensitized solar cells. *Nat. Commun.* **2015**, *6*, 8834–8842.

146. Li, G.; Yella, A.; Brown, D.G.; Gorelsky, S.I.; Nazeeruddin, M.K.; Grätzel, M.; Berlinguette, C.P.; Shatruk, M. Near-IR photoresponse of ruthenium dipyrrinate terpyridine sensitizers in the dye-sensitized solar cells. *Inorg. Chem.* **2014**, *53*, 5417–5419.

147. Swetha, T.; Niveditha, S.; Bhanuprakash, K.; Islam, A.; Han, L.; Bedja, I.M.; Fallahpour, R.; Singh, S.P. New heteroleptic benzimidazole functionalized Ru-sensitizer showing the highest efficiency for dye-sensitized solar cells. *Inorg. Chem. Commun.* **2015**, *51*, 61–65.

148. Argazzi, R.; Larramona, G.; Contado, C.; Bignozzi, C.A. Preparation and photoelectrochemical characterization of a red sensitive osmium complex containing 4,4′,4″-tricarboxy-2,2′:6′,2″-terpyridine and cyanide ligands. *J. Photochem. Photobiol., A* **2004**, *164*, 15–21.

149. Argazzi, R.; Murakami Iha, N.Y.; Zabri, H.; Odobel, F.; Bignozzi, C.A. Design of molecular dyes for application in photoelectrochemical and electrochromic devices based on nanocrystalline metal oxide semiconductors. *Coord. Chem. Rev.* **2004**, *248*, 1299–1316.

150. Altobello, S.; Argazzi, R.; Caramori, S.; Contado, C.; Da Fré, S.; Rubino, P.; Choné, C.; Larramona, G.; Bignozzi, C.A. Sensitization of nanocrystalline TiO$_2$ with black absorbers based on Os and Ru polypyridine complexes. *J. Am. Chem. Soc* **2005**, *127*, 15342–15343.

151. Lapides, A.M.; Ashford, D.L.; Hanson, K.; Torelli, D.A.; Templeton, J.L.; Meyer, T.J. Stabilization of a ruthenium(II) polypyridyl dye on nanocrystalline TiO$_2$ by an electropolymerized overlayer. *J. Am. Chem. Soc.* **2013**, *135*, 15450–15458.

152. Duchanois, T.; Etienne, T.; Cebrián, C.; Liu, L.; Monari, A.; Beley, M.; Assfeld, X.; Haacke, S.; Gros, P.C. An Iron-Based Photosensitizer with Extended Excited-State Lifetime: Photophysical and Photovoltaic Properties. *Eur. J. Inorg. Chem.* **2015**, *2015*, 2469–2477.

153. Kwok, E.C.-H.; Chan, M.-Y.; Wong, K.M.-C.; Lam, W.H.; Yam, V.W.-W. Functionalized alkynylplatinum(II) polypyridyl complexes for use as sensitizers in dye-sensitized solar cells. *Chem. - Eur. J.* **2010**, *16*, 12244–12254.

154. Shinpuku, Y.; Inui, F.; Nakai, M.; Nakabayashi, Y. Synthesis and characterization of novel cyclometalated iridium(III) complexes for nanocrystalline TiO$_2$-based dye-sensitized solar cells. *J. Photochem. Photobiol., A* **2011**, *222*, 203–209.

155. Bozic-Weber, B.; Constable, E.C.; Hostettler, N.; Housecroft, C.E.; Schmitt, R.; Schönhofer, E. The d10 route to dye-sensitized solar cells: step-wise assembly of zinc(II) photosensitizers on TiO$_2$ surfaces. *Chem. Commun.* **2012**, *48*, 5727–5729.

156. Hostettler, N.; Fürer, S.O.; Bozic-Weber, B.; Constable, E.C.; Housecroft, C.E. Alkyl chain-functionalized hole-transporting domains in zinc(II) dye-sensitized solar cells. *Dyes Pigm.* **2015**, *116*, 124–130.

157. Hostettler, N.; Wright, I.A.; Bozic-Weber, B.; Constable, E.C.; Housecroft, C.E. Dye-sensitized solar cells with hole-stabilizing surfaces: "inorganic" versus "organic" strategies. *RSC Adv.* **2015**, *5*, 37906–37915.

158. Housecroft, C.E.; Constable, E.C. The emergence of copper(I)-based dye sensitized solar cells. *Chem. Soc. Rev.* **2015**, *44*, 8386–8398.

159. Bozic-Weber, B.; Constable, E.C.; Housecroft, C.E. Light harvesting with Earth abundant d-block metals: Development of sensitizers in dye-sensitized solar cells (DSCs). *Coord. Chem. Rev.* **2013**, *257*, 3089–3106.

160. Odobel, F.; Pellegrin, Y. Recent Advances in the Sensitization of Wide-Band-Gap Nanostructured p-Type Semiconductors. Photovoltaic and Photocatalytic Applications. *J. Phys. Chem. Lett.* **2013**, *4*, 2551–2564.

161. Ji, Z.; Wu, Y. Photoinduced Electron Transfer Dynamics of Cyclometalated Ruthenium (II)–Naphthalenediimide Dyad at NiO Photocathode. *J. Phys. Chem. C* **2013**, *117*, 18315–18324.

162. Sariola-Leikas, E.; Ahmed, Z.; Vivo, P.; Ojanperä, A.; Lahtonen, K.; Saari, J.; Valden, M.; Lemmetyinen, H.; Efimov, A. Color Bricks: Building Highly Organized and Strongly Absorbing Multicomponent Arrays of Terpyridyl Perylenes on Metal Oxide Surfaces. *Chem.-Eur. J.* **2015**, *22*, 1501–1510.

61

163. Constable, E.C.; Housecroft, C.E.; Šmídková, M.; Zampese, J.A. Phosphonate-functionalized heteroleptic ruthenium(II) bis(2,2′:6′,2″-terpyridine) complexes. *Can. J. Chem.* **2014**, *92*, 724–730.

164. Wood, C.J.; Robson, K.C.D.; Elliott, P.I.P.; Berlinguette, C.P.; Gibson, E.A. Novel triphenylamine-modified ruthenium(II) terpyridine complexes for nickel oxide-based cathodic dye-sensitized solar cells. *RSC Adv.* **2014**, *4*, 5782–5791.

165. Ogura, R.Y.; Nakane, S.; Morooka, M.; Orihashi, M.; Suzuki, Y.; Noda, K. High-performance dye-sensitized solar cell with a multiple dye system. *Appl. Phys. Lett.* **2009**, *94*, 073308.

166. Ozawa, H.; Shimizu, R.; Arakawa, H. Significant improvement in the conversion efficiency of black-dye-based dye-sensitized solar cells by cosensitization with organic dye. *RSC Adv.* **2012**, *2*, 3198–3200.

167. Sharma, G.D.; Daphnomili, D.; Gupta, K.S.V.; Gayathri, T.; Singh, S.P.; Angaridis, P.A.; Kitsopoulos, T.N.; Tasis, D.; Coutsolelos, A.G. Enhancement of power conversion efficiency of dye-sensitized solar cells by co-sensitization of zinc-porphyrin and thiocyanate-free ruthenium(II)-terpyridine dyes and graphene modified TiO_2 photoanode. *RSC Adv.* **2013**, *3*, 22412–22420.

168. Bahreman, A.; Cuello-Garibo, J.-A.; Bonnet, S. Yellow-light sensitization of a ligand photosubstitution reaction in a ruthenium polypyridyl complex covalently bound to a rhodamine dye. *Dalton Trans.* **2014**, *43*, 4494–4505.

169. Koussi-Daoud, S.; Schaming, D.; Fillaud, L.; Trippé-Allard, G.; Lafolet, F.; Polanski, E.; Nonomura, K.; Vlachopoulos, N.; Hagfeldt, A.; Lacroix, J.-C. 3,4-Ethylenedioxythiophene-based cobalt complex: an efficient co-mediator in dye-sensitized solar cells with poly(3,4-ethylenedioxythiophene) counter-electrode. *Electrochim. Acta* **2015**, *179*, 237–240.

Spectroscopic Ellipsometry Studies of *n-i-p* Hydrogenated Amorphous Silicon Based Photovoltaic Devices

Laxmi Karki Gautam, Maxwell M. Junda, Hamna F. Haneef, Robert W. Collins and Nikolas J. Podraza

Abstract: Optimization of thin film photovoltaics (PV) relies on characterizing the optoelectronic and structural properties of each layer and correlating these properties with device performance. Growth evolution diagrams have been used to guide production of materials with good optoelectronic properties in the full hydrogenated amorphous silicon (a-Si:H) PV device configuration. The nucleation and evolution of crystallites forming from the amorphous phase were studied using *in situ* near-infrared to ultraviolet spectroscopic ellipsometry during growth of films prepared as a function of hydrogen to reactive gas flow ratio $R = [H_2]/[SiH_4]$. In conjunction with higher photon energy measurements, the presence and relative absorption strength of silicon-hydrogen infrared modes were measured by infrared extended ellipsometry measurements to gain insight into chemical bonding. Structural and optical models have been developed for the back reflector (BR) structure consisting of sputtered undoped zinc oxide (ZnO) on top of silver (Ag) coated glass substrates. Characterization of the free-carrier absorption properties in Ag and the ZnO + Ag interface as well as phonon modes in ZnO were also studied by spectroscopic ellipsometry. Measurements ranging from 0.04 to 5 eV were used to extract layer thicknesses, composition, and optical response in the form of complex dielectric function spectra ($\varepsilon = \varepsilon_1 + i\varepsilon_2$) for Ag, ZnO, the ZnO + Ag interface, and undoped a-Si:H layer in a substrate *n-i-p* a-Si:H based PV device structure.

Reprinted from *Materials*. Cite as: Gautam, L.K.; Junda, M.M.; Haneef, H.F.; Collins, R.W.; Podraza, N.J. Spectroscopic Ellipsometry Studies of *n-i-p* Hydrogenated Amorphous Silicon Based Photovoltaic Devices. *Materials* **2016**, *9*, 128.

1. Introduction

Development of thin film technologies based on amorphous silicon and germanium, including photovoltaic (PV) devices, involves understanding of material electrical and optical properties [1–4]. It is essential to measure, monitor, and control the thickness, structure, phase, and composition of solar cell component layers in the same configuration used in manufacturing, especially for devices processed over large areas. Doped (*n*-type and *p*-type) and undoped (intrinsic) hydrogenated silicon (Si:H) thin films are used in single, tandem, and multijunction solar cell applications in both *n-i-p* substrate and *p-i-n* superstrate configurations [5–10]. These Si:H films prepared by plasma enhanced chemical vapor deposition (PECVD) may exhibit several structural transitions during growth in the PV device configuration. The structural evolution of Si:H can be controlled by dilution of a reactive silicon carrying gases like silane (SiH_4) with hydrogen (H_2) during deposition. The microstructural evolution of Si:H has been studied for layers deposited on different bulk and thin film substrates with varying degrees of surface roughness, including native and thermal oxide coated crystalline silicon, glass, polyethylene naphthalate (PEN) polymer, as well as underlying structurally distinct Si:H layers prepared under different deposition conditions. A primary technique for studying this growth evolution is the use of near infrared (IR) to ultraviolet (UV) *in situ*, real time spectroscopic ellipsometry (RTSE) applied during Si:H thin film growth [2,4,7,11,12]. RTSE involves collection of ellipsometric spectra as a function of time during a process such as thin film deposition, typically using a multichannel instrument which collects all photon energies in parallel with serial readout of pixels arranged in a one dimensional (1-D) detector. The data acquisition time is typically chosen such that highest signal-to-noise is obtained via averaging of multiple optical cycles over a period of time in which typically only 0.1 to 10 Å of material accumulates. Koh *et al.* [7] reported the growth evolution of intrinsic Si:H on native oxide covered crystalline silicon, amorphous Si:H (a-Si:H) films prepared without additional hydrogen dilution, and on newly deposited ~200 Å p-type microcrystalline or nanocrystalline Si:H (nc-Si:H). These results demonstrate that the nature of the underlying material influences nucleation of crystallites, suppressing nucleation with underlying a-Si:H and promoting nucleation with underlying nc-Si:H. The growth evolution of *p*-layers on specular zinc oxide (ZnO) coated glass and the ability to promote a high nucleation density of nc-Si:H were studied by Rovira *et al.* [11]. In another study of the growth evolution of *p*-type Si:H on ZnO coated glass and ZnO over-coating tin oxide (SnO_2), Koval *et al.* [12] reported that valid material properties and device performance correlations can be better realized for any given material when the properties are obtained from deposition on similar substrates with similar thicknesses as those used in the respective device. The generation of so-called "deposition phase diagrams" or "growth evolution diagrams" for vhf and rf PECVD of Si:H films determined that vhf PECVD shows significant differences in structural evolution with processing conditions, namely the plasma excitation frequency [13]. Growth evolution diagrams have been developed by Stoke *et al.* for intrinsic a-Si:H, amorphous silicon germanium alloys, and nc-Si:H for top, middle, and bottom cell *i*-layers used in triple junction devices [14]. Dahal *et al.* reported growth evolution diagrams for intrinsic and *p*-type Si:H deposited on unoptimized *n*-layer/ZnO/Ag back reflector (BR) coated PEN in *n-i-p* configuration PV devices [15]. Overall these results and analysis procedures developed here are applicable to more directly relating properties of layers in the device configuration, as obtained by non-destructive measurements, with variations in device performance. These types of measurements have been demonstrated for a-Si:H solar cells deposited on planar substrates as described here and also on those incorporating macroscopic roughness or texturing [16–18].

Both a-Si:H and nc-Si:H component materials are used in state-of-the-art Si:H based PV devices. nc-Si:H, either as individual layers or in PV junctions, has significant enhancement in near IR absorption of the solar spectrum and high stability under prolonged illumination in contrast to its amorphous counterpart [19,20]. The quantitative analysis, characterization, and control of the relative nanocrystalline and amorphous volume fractions within mixed-phase films is also a major challenge in Si:H manufacturing. Most often the nanocrystalline fraction is estimated from *x*-ray diffraction or Raman spectroscopy, which can yield values ranging an order of magnitude [6,21,22]. Although these measurements are valuable, limitations exist. Typically *ex situ* *x*-ray diffraction measurements average information over the full depth of a thin film sample, and *ex situ* Raman spectroscopy averages information over a finite penetration depth into the sample that is dependent upon the wavelength of the probing laser, its power, and the absorption coefficient of the material. Profiling these materials non-invasively is an even greater challenge due to probe penetration depth limitations and likely non-uniform crystallite fraction with depth into films. Si:H films may be inhomogeneous with thickness as crystallites nucleate from and coexist with the amorphous phase. Deconvolving gradients in crystallinity from *ex situ* *x*-ray diffraction and Raman spectroscopy measurements requires multiple samples, while *in situ* RTSE measurements applied during film deposition have been used to quantify structural gradients in crystallinity within a single film.

A wealth of information can be extracted from these types of RTSE measurements applied at a single spot on a sample surface, but additional property variations related to sample non-uniformity have also been obtained by *ex situ* mapping spectroscopic ellipsometry (SE) [23–26]. In mapping SE, the sample and multichannel ellipsometer, similar to the instrument used in RTSE studies, are mechanically translated with respect to the each other in one or more dimensions to obtain ellipsometric spectra as a function of spatial position. In the case of Si:H, simplified structural models based on results from RTSE measurements are applied to probe subtle variations in material opto-electronic response such as the band gap of a-Si:H, film thickness, surface roughness thickness, and nanocrystallite fraction in mixed phase materials. These types of measurements have been applied to Si:H [23,24], cadmium telluride [26], and copper indium gallium diselenide [25] PV devices ranging from tens of square centimeters on the laboratory scale to full

65

industrially prepared panels. Thus, improvements in understanding and quantifying the structural transition of Si:H from amorphous to nanocrystalline, as obtained from single spot *in situ* RTSE measurements, can be applied to develop more advanced optical models and more thoroughly analyze mapping SE measurements collected over larger areas.

We have applied SE from 0.734 to 5.88 eV to extract layer thicknesses, interface composition, and optical response in the form of complex dielectric function spectra ($\varepsilon = \varepsilon_1 + i\varepsilon_2$) for all Ag, ZnO, and doped and undoped PECVD Si:H layers found in substrate *n-i-p* PV devices. These studies begin with characterization of ZnO/Ag BRs and have been applied over the near IR to UV spectral range [27,28]. The purpose of BR structures is to increase the optical path length of light within the absorber layer of the PV device, where each photon absorbed has the potential to generate electron-hole pairs and thus electrical current. Any light not absorbed in the first pass of light through the absorber layer is reflected or scattered by the BR back into the absorber layer. Thus PV absorber layers can be made thinner or from materials with low minority carrier diffusion lengths. Due to the substrate dependence of Si:H growth, the same ZnO/Ag BR structures were used to study the growth evolution of doped and undoped Si:H required for use in *n-i-p* a-Si:H PV devices. RTSE using a global $\Sigma\sigma$-minimization analysis procedure has been used to track the behavior of structural transitions in Si:H deposited in the *n-i-p* PV device structure, as functions of hydrogen to reactive gas flow ratio $R = [H_2]/[SiH_4]$, to produce growth evolution diagrams for undoped, *p*-type, and *n*-type layers. Global $\Sigma\sigma$-minimization analysis of RTSE involves using test structural parameters, most commonly a bulk layer thickness (d_b) and surface roughness thickness (d_s), to numerically solve for test ε [29]. These test values of ε are then used to fit other ellipsometric spectra collected at different times when the film is relatively homogeneous. The approach is iterated in order to obtain numerically inverted ε yielding the lowest spectrally and time-averaged error, σ, over the multiple time measurements selected. The numerically inverted ε that minimizes σ are then used to determine structural parameter variations over the full set of RTSE data, with material transitions identified either in the structural parameters themselves or by increases in the error function. Virtual interface analysis (VIA) [30,31] has similarly been applied to track the depth profile of nc-Si:H as well as the formation and stabilization of voids throughout intrinsic layer nc-Si:H growth. In VIA, the full sample stack is not analyzed. Instead, optical properties of a pseudo-substrate are generated from ellipsometric spectra collected earlier in the deposition by numerically inverting the measurement to obtain ε using a simplified model consisting of a semi-infinite pseudo-substrate and a surface roughness layer. The effective ε for the pseudo-substrate contains information of all underlying material(s) in the sample stack and is then used as the semi-infinite substrate for analysis of subsequent data sets. In this sense, the time derivative of ellipsometric spectra is analyzed and full understanding of the underlying structure is not required in the analysis procedure. When combined with Σσ-minimization approaches for structurally graded Si:H, VIA yields ε for both the nc-Si:H and a-Si:H components as well as the time and bulk layer thickness dependence of component material fractions in the overlayer of material accumulated between each pseudo-substrate and subsequent data set pair. These techniques are used to provide guidance for the deposition and *in situ* characterization of a-Si:H, nc-Si:H, and mixed-phase (a+nc)-Si:H layers during growth in device structures.

In addition to RTSE studies of material growth evolution limited to the near IR to UV spectral range, we also have used *ex situ*, room temperature IR-extended SE (IR-SE) from 0.04 to 0.75 eV using a Fourier transform IR ellipsometer operating over this spectral range. This extension enables spectra in ε for Si:H and BR components layers to be determined from the mid-IR to UV wavelength range. Simultaneous analysis of ellipsometric spectra collected from multiple samples consisting of BR and *i*-layer/*n*-layer/BR stacks deposited on borosilicate glass substrates was used to yield a common ε for each layer while structural parameters such as d_b and d_s may be varied separately as in RTSE data analysis [4,7,11,13–15,27,28] and similar in methodology to the divided spectral range approach [32]. A common parameterization of ε is used to fit *ex situ* ellipsometric spectra collected from separate near IR-UV multichannel and FTIR ellipsometers. This approach yields a continuous set of spectra in ε for each material, although the particular beam spot location on the sample surface may not be the same during measurement using each ellipsometer [33]. The results of the analysis of IR-SE data were used to study absorption in the BR components, which can be used to extract electrical transport properties and phonon modes. In addition, εfor protocrystalline intrinsic a-Si:H extracted from IR-SE data is sensitive to the silicon-hydrogen bonding configuration. Comparison of optical absorption features affords a method of assessing film structural and chemical character, which then suggests ways to improve material quality and potentially device performance [34,35]. PV devices incorporating optimization principles based on IR spectroscopy and RTSE analyzed for each layer in the device configuration have exhibited relatively high performance [23,34].

In situ RTSE studies have been used to yield growth evolution phase diagrams of each doped and undoped Si:H layer in the *n-i-p* PV device configuration and structural evolution profiles of crystallite and void fractions. The information from growth evolution diagrams was used to design PV device structures, lacking the *p*-layer and incorporating only the amorphous phase, for characterization using *ex situ* IR-extended SE. Results of *ex situ* IR-extended SE combined with *ex situ* near IR-UV SE have been used to obtain spectra in ε from 0.04 to 5.0 eV for ZnO and a-Si:H in the BR/*n-i-p* a-Si:H solar cell configuration. Higher energy transitions related to electronic structure in Ag, ZnO, and a-Si:H; the band gap in ZnO and a-Si:H; a plasmon feature in the Ag + ZnO interface; IR vibrational modes related to chemical bonding in a-Si:H and ZnO; and free carrier absorption in Ag and the Ag + ZnO interface have been obtained from spectra in ε. In addition to information on each of these materials in the *n-i-p* device structure, the structural and optical properties derived here can be applied in the future analysis of *ex situ* SE in either single spot [36] or mapping configurations.

2. Experimental Details

Thin film doped and undoped Si:H films were deposited using a load-locked rf (13.56 MHz) PECVD reactor onto 6″ × 6″ borosilicate glass substrates coated with rf magnetron sputtered ZnO/Ag BRs as are commonly used in *n-i-p* configuration solar cells. Si:H *n-*, *p-*, and *i*-layers have been prepared using different hydrogen dilution ratios, *R*, onto BR coated substrates or those otherwise mimicking the outermost previous layer in the device structure to generate growth evolution diagrams and identify the optimum conditions for protocrystalline a-Si:H for solar cells. Table 1 lists the deposition parameters used for fabrication of each layer including gas flows, pressure (*p*), power density (*P*), and substrate temperature (*T*). The deposition conditions used here were adopted from the Dahal *et al.*, 2013 and Dahal *et al.*, 2014 [23,37]. The deposition was done in the same chamber corresponding to reasonable device quality materials as evidenced by incorporation in *n-i-p* solar cells without textured BRs yielding ~7.5% efficiency. Maximum device performance parameters are open circuit voltage of 0.90 V, short circuit current of 12.5 mA/cm^2, and fill factor of 70%. The open circuit voltage and fill factor are reasonable for moderate quality *n-i-p* solar cells. The short circuit current is low as the specular BR does not produce the level of scattering expected from a textured BR where the optical path length of long wavelength light not initially absorbed in the intrinsic a-Si:H layer is increased. This increase in optical path length results in increased current density, which is absent for specular devices. The *n*-type Si:H films were deposited onto BR coated borosilicate glass. The intrinsic layers were deposited onto BR's coated with *n*-type a-Si:H prepared at *R* = 50. The *p*-type Si:H films were deposited onto borosilicate glass initially coated with ~3000 Å thick intrinsic a-Si:H prepared at *R* = 10. This intrinsic layer is deposited to eliminate any contributions to the microstructural evolution from the underlying glass substrate. Variable parameters were *R* = [H$_2$]/[SiH$_4$] for all three layers. The dopant gas ratios for *n-* (*D* = [PH$_3$]/[SiH$_4$]) and *p*-layers (*D* = [B$_2$H$_6$]/[SiH$_4$]), which can have significant influence on structural and the electronic properties, were fixed at *D* = 0.0125. Cr, Ag, and ZnO layers were prepared by rf magnetron sputtering at room temperature. Here, Cr was used as an adhesion interlayer to avoid delamination of the Ag film from the borosilicate glass. The ZnO/Ag BR structures were prepared under identical conditions for each sample to study how the Si:H layers grow in the device configuration.

Table 1. Deposition conditions for the individual layers in the a-Si:H *n-i-p* solar cell configuration deposited on 6″ × 6″ borosilicate glass substrates. The 5% dopant gas in H_2 is by volume. Cr, Ag, and ZnO were sputtered at room temperature (RT).

Layer	Substrate Temperature T (°C)	Pressure p (mTorr)	Radio Frequency (rf) Power P (W/cm²)	Gas Flow (sccm)				
				Ar	SiH₄	5% PH₃ or B₂H₆ in H₂	H₂	R = [H₂]/[SiH₄]
Cr	RT	5	0.92	10	-	-	-	-
Ag	RT	5	0.92	10	-	-	-	-
ZnO	RT	5	0.92	10	-	-	-	-
n	200	1500	0.032	-	2	0.5 PH₃	40–160	20–80
i	200	800	0.04	-	5	-	50–250	10–50
p	100	1500	0.066	-	2	0.5 B₂H₆	100–400	50–200

RTSE was performed *in situ* at a single spot during deposition using a rotating-compensator multichannel ellipsometer (J. A. Woollam Company model M-2000) that can measure ellipsometric spectra (in the form of $N = \cos 2\psi$, $C = \sin 2\psi \cos \Delta$, $S = \sin 2\psi \sin \Delta$) from 0.734 to 5.88 eV with a minimum data acquisition time of 50 ms [38,39]. This type of instrument collects ellipsometric spectra at all photon energies in parallel by a combination of a 1-D linear detector array and serial pixel readout. Dual detectors are required to access this spectral range, and consist of a silicon based charged coupled device (CCD) and indium gallium arsenide photodiode array (PDA). RTSE measurements were collected at the respective deposition temperature at angles of incidence near 70° and spectra obtained from single optical cycles were averaged over 1.5 s intervals to increase the signal-to-noise ratio. Analysis of experimentally collected RTSE data was performed using J. A. Woollam Co. CompleteEASE software (Lincoln, NE, USA). The time evolution of d_b and d_s as well as the spectroscopic ε of the bulk Si:H layers were extracted from RTSE data using a global $\Sigma\sigma$-minimization procedure. For Si:H films, global $\Sigma\sigma$-minimization analysis of RTSE involves using test d_b and d_s values for the Si:H layer being deposited on top of a pre-defined substrate stack to numerically solve for test ε of the Si:H layer [29]. The test values of ε are then used to fit other spectra collected at different times when the film is relatively homogeneous, typically near 100–200 Å in accumulated material thickness for Si:H films where structural transitions have not yet had time to mature. The approach is applied in the regime prior to crystallite nucleation and is iterated in order to obtain numerically inverted ε yielding the lowest spectrally and time-averaged error, σ, over the multiple time measurements selected. The numerically inverted ε minimizing σ are taken to be the best representation of the a-Si:H optical properties and are then used to determine structural parameter variations over the full set of RTSE data, with the nucleation of crystallites from the a-Si:H matrix identified by a sharp increase in the surface roughness thickness ($\geqslant 1$ Å between successive time points) and by increases in σ as spectra in ε for a-Si:H are no longer adequate to fit ellipsometric spectra collected as nanocrystallites evolve. The unweighted error function, σ, is defined by [40]:

$$\sigma = \sqrt{\frac{1}{3N-M} \sum_{j=1}^{N} \left[\begin{array}{l} \left(\cos2\psi_j^{\mathrm{mod}} - \cos2\psi_j^{\mathrm{exp}}\right)^2 \\ + \left(\sin2\psi_j^{\mathrm{mod}}\cos\Delta_j^{\mathrm{mod}} - \sin2\psi_j^{\mathrm{exp}}\cos\Delta_j^{\mathrm{exp}}\right)^2 \\ + \left(\sin2\psi_j^{\mathrm{mod}}\sin\Delta_j^{\mathrm{mod}} - \sin2\psi_j^{\mathrm{exp}}\sin\Delta_j^{\mathrm{exp}}\right)^2 \end{array} \right]} \tag{1}$$

where N is the number of measured values; and M the number of fit parameters; "*exp*" denotes experimental spectra; and "*mod*" denotes that generated from the model. An advantage of conducting *in situ* RTSE measurements during growth of Si:H by PECVD is that spectra in ε can be obtained prior to the exposure of the sample to ambient and potential oxidation.

Near IR to near UV room temperature ellipsometric spectra over a range from 0.734 to 5.88 eV were collected at a single spot *ex situ* prior to the collection of the *ex situ* IR-SE data as those measurements were not able to be collected *in situ* during film growth. IR-SE data was collected at a single spot using a similar single rotating compensator instrument (J. A. Woollam Company, Lincoln, NE, USA, model FTIR-VASE) from 0.04 to 0.75 eV at 1 cm^{-1} resolution [41]. The angle of incidence for all *ex situ* measurements was nominally 70°. Ellipsometric spectra over the mid-IR to near UV range collected from the two instruments were analyzed simultaneously using a common parameterization for ε based on structural models initially developed from RTSE and near IR to near UV measurements and J. A. Woollam Co. WVASE software (Lincoln, NE, USA). The error function in Equation 1 is also used for analysis and fitting of *ex situ* IR-SE data.

For *in situ* RTSE, *ex situ* near IR to UV SE, and *ex situ* IR-extended SE data analysis, the optical response of the surface roughness layer of thickness d_s for Si:H and ZnO is represented using Bruggeman effective [42,43] medium approximation mathematically represented as:

$$\sum_n f_n \frac{\varepsilon_n - \varepsilon}{\varepsilon_n + 2\varepsilon} = 0. \tag{2}$$

in this expression, material fractions (f_n) and component material optical response (ε_n) are used to generate a composite ε for the mixture. For surface roughness in this work, spectra in ε from Bruggeman effective medium approximation consist of 0.5 void and 0.5 underlying material volume fractions, regardless of composition of the underlying layer.

The structural model constructed for each sample consists of a stratified layer stack of optically distinct materials, which may be continuous films, interfaces with unique ε and thickness, or Bruggeman effective medium approximation layers. Ellipsometric spectra are modeled using a scattering matrix formalism [44] in which 2×2 matrices based on Fresnel coefficients and wave propagation of light through media is generated for the semi-infinite substrate and ambient, each layer of finite thickness, and the interfaces between each optically distinct layer. Matrices are calculated for the components of the incident electric field both parallel and perpendicular to the plane of incidence. In general, ellipsometric spectra are sensitive to spectra in ε for each material, including the ambient and substrate, and the thicknesses of optically finite layers.

3. Results and Discussion

3.1. Optical Characterization of Back Reflector Components and Structure

The first layers deposited for *n-i-p* configuration a-Si:H solar cells comprise the ZnO/Ag BR structure. Therefore, ellipsometric spectra from 0.04 to 5.0 eV are collected and analyzed for a ZnO/Ag BR structure. The models and thicknesses described here first correspond to a ZnO coated Ag BR sample, while variations in properties due to growth of over-deposited *n*- and *i*-type a-Si:H layers is described in Section 3.3.1. All layers were deposited without vacuum break with conditions given in Table 1. *In situ* SE data from 0.734 eV to 5.88 eV was collected for each deposited layer and the model generated was used for extended spectral range IR-SE analysis.

3.1.1. Ag and ZnO + Ag Interface Properties

Data collected for semi-infinite Ag substrate were taken before ZnO layer deposition at room temperature and were analyzed in the energy range from 0.734 to 5.88 eV. Figure 1 shows spectra in ε for Ag parameterized by a combination of a Drude oscillator [45], a higher energy transition assuming critical point parabolic bands (CPPB) [46], and a constant additive term to ε_1 denoted as ε_∞. The surface roughness is represented by two Lorentz oscillators [47] with $\varepsilon_\infty = 1$. The Drude oscillator is represented by:

$$\varepsilon(E) = \frac{-\hbar^2}{\varepsilon_0 \rho (\tau E^2 + i\hbar E)} \tag{3}$$

where $\hbar$ is the reduced Planck's constant, ε_0 is the permittivity of free space, τ is the scattering time, and ρ is the resistivity. Each CPPB oscillator is represented by:

$$\varepsilon(E) = A e^{i\phi} \left\{ \frac{\Gamma}{[2E_n - 2E - i\Gamma]} \right\}^\mu \tag{4}$$

where A, E_n, Γ, μ, and φ are the amplitude, resonance energy, broadening, exponent, and phase of the critical point, respectively. The exponent μ can assume the values of $1/2, 0$, and $-1/2$ depending on whether the critical points are one, two, or three dimensional in nature. In this work, only the one dimensional CPPB oscillator has been used, and so its value was fixed at $\mu = 0.5$. Each Lorentz oscillator is represented by:

$$\varepsilon(E) = \frac{A\Gamma E_0}{\left[E_0{}^2 - E^2 - i\Gamma E\right]} \qquad (5)$$

where A, Γ, and E_0 represent amplitude, broadening, and resonance energy respectively. All parameters describing Ag and its surface roughness are listed in Table 2. A resistivity of $3.02 \pm 0.03 \times 10^{-6}$ Ωcm and a scattering time of 16.7 ± 0.1 fs were determined from the Drude oscillator parameters of the Ag film.

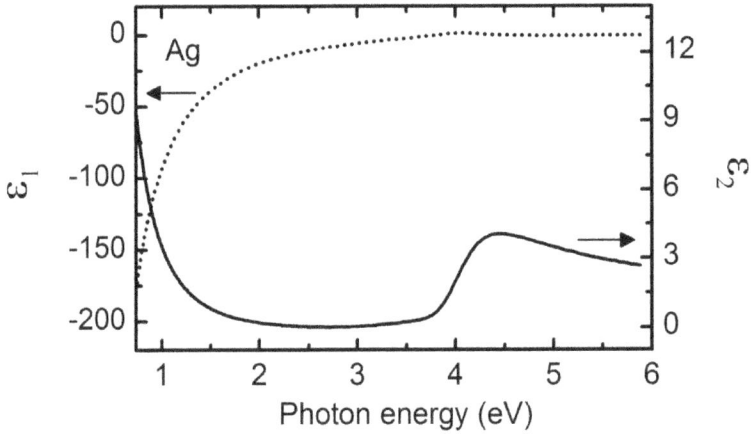

Figure 1. Complex dielectric function spectra, $\varepsilon = \varepsilon_1 + i\varepsilon_2$, (arrow pointing left for ε_1 axis, arrow pointing right for ε_2 axis) from 0.734 to 5.88 eV for a semi-infinite Ag film parameterized with a combination of a Drude oscillator and two oscillators assuming critical point parabolic bands (CPPB) with parameters listed in Table 2.

Table 2. Parameters describing complex dielectric function ($\varepsilon = \varepsilon_1 + i\varepsilon_2$) and structure for a semi-infinite Ag film on a borosilicate glass over coated by Cr before ZnO deposition. Experimental ellipsometric spectra were collected *in situ* after deposition at room temperature in the spectral range from 0.734 to 5.88 eV and fit using least square regression analysis with an unweighted estimator error function, $\sigma = 5 \times 10^{-3}$. For bulk Ag, the parameterization of ε consisted of a Drude oscillator, two oscillators assuming critical point parabolic bands (CPPB), and a constant additive term to ε_1 denoted ε_∞. Spectra in ε for the 30 ± 2 Å surface roughness layer were parameterized with two Lorentz oscillators and $\varepsilon_\infty = 1$.

Ag Surface Roughness					
Oscillator	A (Unitless)	Γ (eV)	E_0 (eV)	-	-
Lorentz	4.2 ± 0.2	2.5 ± 0.1	5.17 ± 0.02	-	-
Lorentz	1.0 ± 0.3	0.06 ± 0.03	3.61 ± 0.01	-	-
Bulk Ag					
Oscillator	A (Unitless)	Γ (eV)	E_n (eV)	Θ (degrees)	μ
CPPB	5.29 ± 0.09	0.70 ± 0.03	3.845 ± 0.008	-180.306 ± 0.002	0.5
CPPB	10.39 ± 0.07	0.87 ± 0.01	4.025 ± 0.001	-7.0 ± 0.4	0.5
Drude	ρ (Ωcm)			τ (fs)	
Constant additive term to ε_1	$3.02 \pm 0.03 \times 10^{-6}$			16.7 ± 0.1	
ε_∞			1.632 ± 0.008		

The structural model for the ZnO/Ag BR in the energy range 0.734 to 5 eV consisted of a semi-infinite Ag metal layer deposited onto glass, a 108 ± 10 Å ZnO + Ag interfacial layer, a 3059 ± 3 Å bulk ZnO layer, and a 80 ± 1 Å surface roughness represented using Bruggeman effective medium approximation of 0.5 ZnO and 0.5 void volume fractions. Parametric expressions were used to describe ε for Ag, ZnO, and the ZnO + Ag interface and are listed in Tables 2 and 3. Previous studies of ZnO/Ag interfaces in the BR of thin film *n-i-p* a-Si:H PV shows that the optically determined value of Ag surface roughness obtained from RTSE is very close to that measured with atomic force microscope (AFM) with $d_{s,RTSE}$ (Å) = 0.96 $d_{s,AFM}$ (Å) + 5 Å [28]. The $d_{s,RTSE} = 30 \pm 2$ Å for Ag corresponds to a $d_{s,AFM} = 26$ Å. After deposition of ZnO, the ZnO/Ag interface layer thickness is reported by Dahal *et al.* as d_i (Å) = 1.98 d_s (Å) + 17.5 Å. The interface layer thickness predicted from the Ag surface roughness in this work is 76.9 Å as compared to that obtained in our parametric analysis of 108 Å [28]. Our parametric value slightly overestimates the prediction, however in Dahal *et al.* [28] the samples with similar Ag surface roughness, 25–30 Å, also has an interface thickness of 75–110 Å which are greater than the linear prediction. Figure 2 shows that the spectra in ε obtained for the ZnO + Ag interface is optically different than Ag and ZnO alone and can be modeled by a Lorentz oscillator and a Drude oscillator in the near IR to near UV range (0.734 to 5 eV) with $\varepsilon_\infty = 1$. The ZnO + Ag interface exhibits a clear localized particle plasmon absorption feature which can be modeled using a Lorentz oscillator with a resonance energy at 2.83 ± 0.01 eV [27,48]. A resistivity of $3.7 \pm 0.5 \times 10^{-5}$ Ωcm and a scattering time of

2.7 ± 0.3 were determined from the Drude oscillator parameters of the ZnO + Ag interface. These values indicate that when compared to bulk Ag, the interface is less conductive due to incorporation of higher resistivity undoped ZnO and potentially more disordered as suggested by the lower scattering time. Over this spectral range, ε for ZnO was initially fit using two CPPB oscillators, ε_∞, and a zero-broadened Sellmeier oscillator [49] represented by:

$$\varepsilon(E) = \frac{A}{(E_n^2 - E^2)} \tag{6}$$

where A and E_n represent the amplitude and resonance energy, respectively.

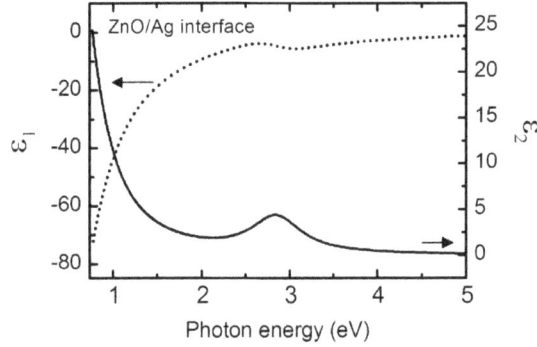

Figure 2. Spectra in ε (arrow pointing left for ε_1 axis, arrow pointing right for ε_2 axis) from 0.734 to 5.0 eV for the 108 ± 1 Å thick ZnO + Ag interface layer parameterized with a Lorentz and a Drude oscillator with parameters listed in Table 3.

3.1.2. Phonon Modes in ZnO

The analysis was extended to the IR by fitting parameters defining ε for ZnO only and fixing those defining ε for Ag and the ZnO + Ag interface as well as the interface layer thickness. This analysis approach was chosen because free carrier absorption represented by the Drude feature dominates the IR response of Ag and the ZnO + Ag interface layers and is already established from near IR to UV spectral range analysis. A common parameterization of ε for the ZnO was applied for the data collected from the two instruments with spectral ranges from 0.04 to 0.734 eV and 0.734 to 5.0 eV, respectively, although the bulk ZnO layer thickness was allowed to vary for the ellipsometric spectra collected from each respective instrument to account for measurement on different spots over the sample surface. A common surface roughness thickness between the two sets of measured spectra was obtained, as this effect will vary less with non-uniformity than the overall bulk layer thickness. Figure 3 shows ε for ZnO represented by a combination of CPPB oscillators for electronic transitions, Lorentz oscillators representing IR phonon modes, and a constant real additive term ε_∞ to account for dispersion from absorption features outside the measured spectral range from 0.04 to 5 eV with parameters given in Table 4. The near IR to near UV range shows only small absorption below the lowest direct transition at 3.364 eV as expected for direct band gap ZnO [50]. Phonon modes for wurtzite ZnO are $\Gamma_{opt} = 1A_1 + 2B_1 + 1E_1 + 2E_2$, with A_1 and E_1 modes IR-active. Only one characteristic transverse optical (TO) mode for ZnO with E_1 symmetry at 0.0501 eV (404.08 cm^{-1}) is resolved for this sample [51–53]. Weak absorption bands in the spectral region from 0.134 to 0.264 eV (1080 to 2130 cm^{-1}) have been observed and are often associated with hydrogen-associated bending modes; stretching modes of hydrogen bonded to heavier elements like zinc; and various carbon, oxygen, and nitrogen-related stretching modes not involving hydrogen [54]. These types of peaks are analogous to those found in the absorbance spectra from traditional unpolarized FTIR measurements, which lack sensitivity to discerning thickness and the full complex optical properties simultaneously—a capability of SE measurements.

Table 3. Parameters describing ε and structure for a ZnO film deposited on Ag and the ZnO + Ag interface formed. Experimental ellipsometric spectra were collected *in situ* after deposition at room temperature in the spectral range from 0.734 to 5.0 eV and fit using least squares regression analysis with an unweighted estimator error function, $\sigma = 7 \times 10^{-3}$. Parameters describing ε for Ag were fixed from Table 2. For ZnO, the parameterization of ε consisted of two CPPB oscillators, a Sellmeier oscillator, and ε_∞. For the ZnO + Ag interface, the parameterization of ε consisted of a Drude oscillator, a Lorentz oscillator, and ε_∞.

Layer	Oscillators			
	CPPB ($\mu = 0.5$)	$\varepsilon_\infty = 2.27 \pm 0.01$		
	A (Unitless)	Γ (eV)	E_n (eV)	Θ (degrees)
ZnO $d_b = 3060 \pm 3$ Å $d_s = 80 \pm 1$ Å	2.63 ± 0.02	0.199 ± 0.002	3.363 ± 0.001	−20.1 ± 0.5
	1.41 ± 0.02	3.83 ± 0.08	4.36 ± 0.03	0 (fixed)
	Sellmeier			
	A (eV^2)	Γ (eV)	E_n (eV)	
	0.080 ± 0.002	-	0	
	Lorentz	$\varepsilon_\infty = 1$		
	A (Unitless)	Γ (eV)	E_0 (eV)	
ZnO/Ag Interface = 108 ± 11 Å	2.8 ± 0.2	0.57 ± 0.05	2.83 ± 0.01	
	Drude			
	ρ (Ω cm)	τ (fs)		
	3.7 ± 0.5 x10^{-5}	2.7 ± 0.3		

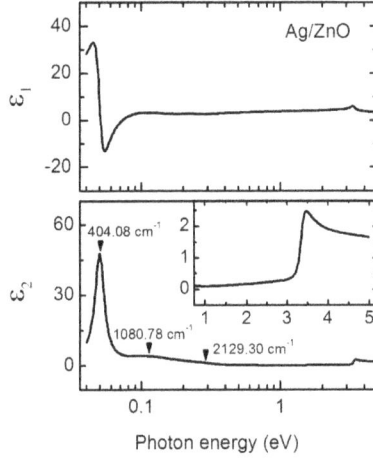

Figure 3. Spectra in ε (top panel, real part ε_1; bottom panel, imaginary part ε_2) from 0.04 to 5.0 eV for a 3010 ± 2 Å thick ZnO film on Ag, with ε for ZnO parameterized using a combination of two CPPB and three Lorentz oscillators with parameters listed in Table 4. The inset shows high-energy electronic transitions in ε_2.

Table 4. Parameters describing ε and structure for a ZnO film deposited in a ZnO/Ag back reflector (BR). Experimental ellipsometric spectra were collected *ex situ* using near infrared to ultraviolet (0.734 to 5.0 eV) and infrared (0.04 to 0.734 eV) spectral range instruments and fit jointly using least squares regression analysis with an unweighted estimator error function, $\sigma = 8 \times 10^{-3}$. Parameters describing ε for Ag and the ZnO + Ag interface were fixed from Tables 2 and 3 respectively. The ZnO bulk layer thickness was allowed to vary separately for each set of ellipsometric spectra; all other parameters are common to both analyses. For ZnO, the parameterization of ε consisted of two CPPB oscillators, three Lorentz oscillators, and ε_∞.

Layer	Oscillators			
	CPPB ($\mu = 0.5$)	$\varepsilon_\infty = 2.43 \pm 0.01$		
	A (Unitless)	Γ (eV)	E_n (eV)	Θ (degrees)
ZnO d_b (Near IR to UV) = 2996 $\pm$ 2 Å d_b (IR) = 3025 $\pm$ 2 Å d_s = 84 $\pm$ 1 Å	2.82 $\pm$ 0.02	0.209 $\pm$ 0.002	3.364 $\pm$ 0.001	-20.8 ± 0.4
	1.23 $\pm$ 0.02	3.95 $\pm$ 0.03	3.94 $\pm$ 0.02	0
	Lorentz			
	0.75 $\pm$ 0.05	0.196 $\pm$ 0.005	0.264 $\pm$ 0.002	-
	3.17 $\pm$ 0.03	0.169 $\pm$ 0.007	0.134 $\pm$ 0.001	-
	46 $\pm$ 2	0.0093 $\pm$ 0.0004	0.0501 $\pm$ 0.0002	-

3.2. RTSE Monitoring of Si:H in n-i-p Solar Cell Devices

The films used to develop growth evolution diagrams for doped and undoped Si:H deposited in the glass substrate/BR/*n-i-p* a-Si:H device configuration were grown as a function of R in an effort to probe the subtle fluctuations expected as the material transitions from amorphous to nanocrystalline [31,55]. A distinct type of roughening transition is reported in which crystallites nucleate from the growing amorphous phase. Because of the low crystallite nucleation density as observed by Fujiwara *et al.* and Ferlauto *et al.* [4,31], the growth of crystalline protrusions produce a roughness layer that increases promptly when compared to increases in bulk layer thickness. Thus, the onset of roughening identifies a transition to mixed-phase amorphous+nanocrystalline (a+nc)-Si:H film growth accompanied by changes in the film optical properties. This behavior denotes the amorphous-to-mixed-phase [a→(a+nc)] transition. Simple roughening of the amorphous phase also tends to exhibit a lower increase in d_s, accompanied by only minimal increases in σ, with accumulated bulk layer thickness. Crystallites nucleating from the amorphous phase grow preferentially over the surrounding material, until the point at which the crystallites cover the surface. The disappearance of the amorphous phase and coalescence of crystallites is denoted as the mixed-phase-to-single-phase nanocrystalline [(a+nc)→nc] transition. The a→(a+nc) and (a+nc)→nc transitions were detected using RTSE monitoring and data analysis in this work. The Si:H films prepared at low R remain in the amorphous growth regime throughout the deposited thickness. VIA was applied to RTSE data collected during growth for Si:H transitioning from amorphous to nanocrystalline [31,56,57].

Figure 4 shows an example of the results of VIA applied to RTSE data to obtain the surface roughness thickness, nanocrystallite fraction, void fraction, and average mean square error (Equation (1)) as functions of the bulk layer thickness for a $R = 50$ *i*-layer on a BR over-coated with a 200 Å $R = 50$ *n*-layer. The VIA applied here utilizes spectra from 2.75 to 5.0 eV and ε for a-Si:H and nc-Si:H components as shown in Figure 5. Spectra in ε for nc-Si:H was obtained from the end of the respective deposition when the film is known to be fully nanocrystalline, ~1150 Å of a 1300 Å thick film using the same optical model as was used for the *i*-layer growth evolution diagram. In this model the free parameters are d_b and d_s. Spectra in ε for the amorphous phase was taken from the analysis of $R = 15$ deposition corresponding to a time within the first ~200 Å of bulk material prior to the nucleation of nanocrystallites. There is strong optical contrast between the two sets of ε for Si:H, in that the amorphous phase has only a single broad resonance while that of nanocrystallite material has two features representative of dampened and broadened critical point features found in single crystal silicon [58]. These reference spectra in ε for a-Si:H and nc-Si:H, along with that for void ($\varepsilon = 1$), were then used in a three component Bruggeman effective medium approximation [42,43] layer and a least-squares regression within the VIA with d_s and the relative nanocrystallite (f_{nc}) and void (f_{void}) fractions as free parameters and the amorphous fraction constrained ($f_a = 1 - f_{nc} - f_{void}$).

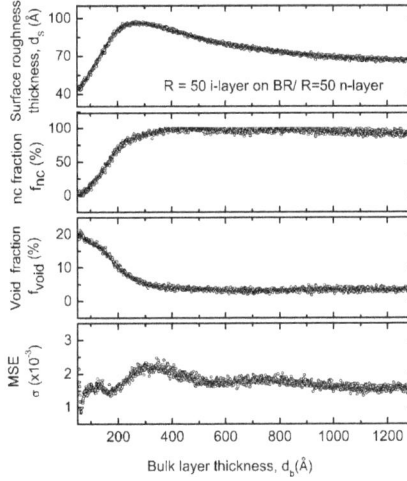

Figure 4. Mean square error (MSE), void fraction (f_{void}), nanocrystalline volume fraction (f_{nc}), and surface roughness thickness (d_s) in the top ~10 Å of the bulk layer, plotted *versus* the accumulated bulk layer thickness for an intrinsic hydrogen diluted $R = [H_2]/[SiH_4] = 50$ Si:H film deposited on a 200 Å $R = 50$ n-type a-Si:H over-deposited onto a ZnO/Ag back reflector (BR), as determined by virtual interface analysis (VIA) applied to real time spectroscopic ellipsometry (RTSE) data. Spectrally averaged mean error for f_{void}, f_{nc}, and d_s are 0.3%, 2.4%, and 0.8 Å respectively.

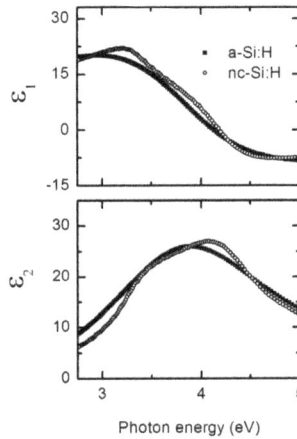

Figure 5. Spectra in ε (top panel, real part ε_1; bottom panel, imaginary part ε_2) of a-Si:H and nc-Si:H reference material used in VIA applied over a spectral range from 2.75 to 5.0 eV. Spectra in ε for a-Si:H and nc-Si:H were obtained from analysis of RTSE data and by numerical inversion at a bulk layer thickness of 200 and 1150 Å, respectively.

Results of VIA show an increase in surface roughness followed by a decrease within the first ~300 Å of material accumulation, indicating crystallite nucleation on the substrate followed by coalesce of the clusters. The nanocrystallite fraction increases with bulk layer thickness, then converges to 1.0 as expected for a nanocrystalline film. Voids initially appear with the nucleation of crystallites, which then subsequently decrease and stabilize near f_{void} = 0.04 throughout the growth of this layer. Depending on the source of reference ε for nc-Si:H, this behavior could indicate that the grains under these conditions were not well passivated with a-Si:H as is desirable in nc-Si:H PV [22,59]. Optimized nanocrystalline/microcrystalline PV devices often incorporate layers prepared at lowest hydrogen dilution where crystallite growth can occur, and nc-Si:H layers are often fabricated using hydrogen dilution grading approaches to manipulate the degree of crystallinity. For very high values of hydrogen dilution, such as R = 50 in this example, the material is likely not optimized for solar cells, because cracks related to voids can promote shunts in the cells and channels by which contamination (e.g., oxygen) can enter into the layer [14,21,60].

Comparison of the structural behavior of the a→(a+nc) and (a+nc)→nc transitions as a function of single deposition parameters has been used to produce so-called deposition phase diagrams or growth evolution diagrams which have helped guide the development of optimization principles in Si:H based PV. For example, the structural evolution can be controlled by the dilution of reactive silicon carrying gases with hydrogen during the deposition process. Films prepared at low R remain amorphous throughout their total thickness, while those prepared at higher R nucleate crystallites. The thickness at which the a→(a+nc) transition occurs decreases with increasing R. Optimum a-Si:H based PV devices incorporate layers prepared at the highest R that will remain amorphous throughout the full thickness of the absorber layer while optimum nc-Si:H PV incorporates layers prepared at the lowest R where crystallite growth can occur [7,59–62]. For the case of a-Si:H, the additional hydrogen dilution improves ordering in the a-Si:H network, while for nc-Si:H low hydrogen dilution ensures that hydrogen etching does not occur and the grain boundaries remain well-passivated.

The growth evolution diagrams of n-type, intrinsic, and p-type Si:H layers in the n-i-p/BR/glass configuration are depicted in Figure 6. The n-type Si:H layers are prepared at T = 200°C, p = 1.5 Torr, P = 0.032 W/cm^2, and D = 0.0125 as a function of R varied from 20 to 80. For R < 50 the n-layer remains amorphous at least to a thickness of 500 Å. At R = 50 nanocrystallites nucleate in the n-type Si:H at about 450 Å of bulk layer thickness. The amorphous material prior to the a→(a+nc) transition of these depositions is protocrystalline [2]. A ~200 Å thick n-layer is typical for n-i-p configuration devices, and the best R for optimized n-i-p a-Si:H solar cells with a protocrystalline n-layer is identified here as near

$R = 50$. As R is further increased, nanocrystallites nucleate within the amorphous phase at decreasingly lower thicknesses as indicated by the a→(a+nc) transition thicknesses. Films nucleating crystallites and grown to sufficient thickness show the (a+nc)→nc transition with crystallites coalescing at similarly decreasing thickness with increasing R. The film at $R = 60$ nucleates crystallites at ~100 Å and coalescence occurs at ~380 Å. These transitions occur much sooner for $R = 80$ leading to nanocrystallite formation in the very beginning of the deposition, making it unsuitable for an optimum n-type layer in single junction a-Si:H devices.

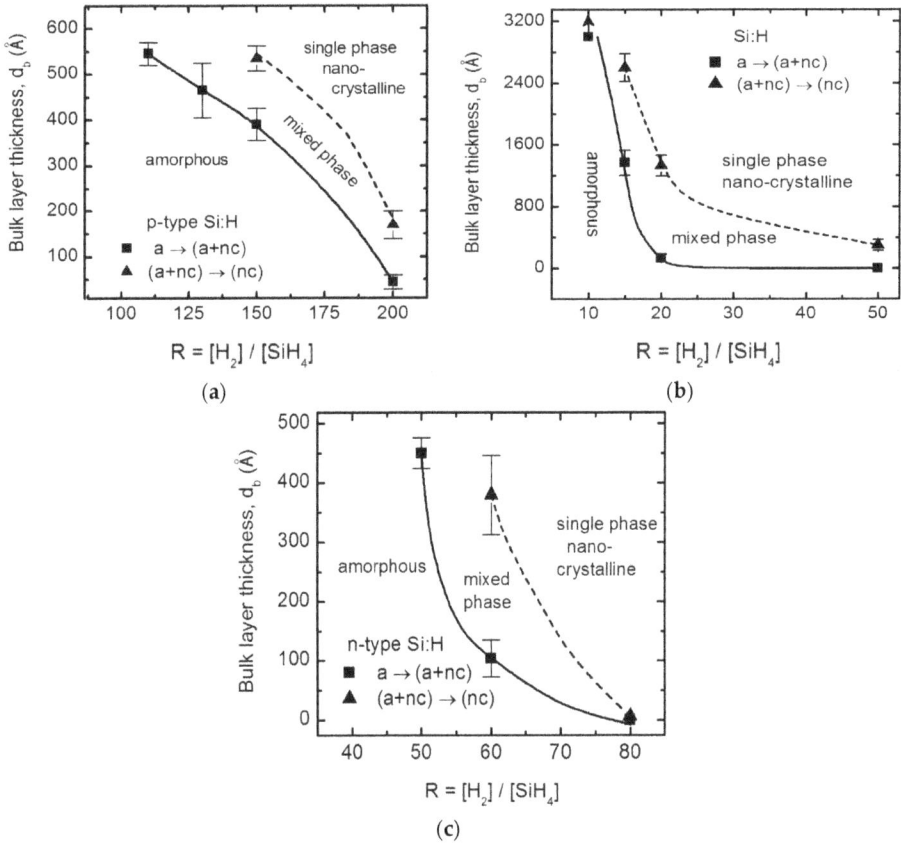

Figure 6. Growth evolution diagrams obtained from analysis of RTSE data for (a) p-type; (b) intrinsic; and (c) n-type Si:H as a function of variable hydrogen dilution $R = [H_2]/[SiH_4]$ in the n-i-p solar cell device structure. The data values and connecting lines depict the a→(a+nc) and (a+nc)→nc structural transitions of doped and undoped Si:H prepared at conditions described in Table 1. Arrows pointing upward indicate the respective transition occurs beyond the maximum thickness measured.

In both *n-i-p* substrate and the *p-i-n* superstrate PV device configurations, most incident photons are absorbed in the intrinsic layer with photo-generated electrons and holes transported to the contacts. Hence, optimization of *i*-layer is critical and the optical response and phase composition of these intrinsic layers tremendously impact solar cell performance. The intrinsic Si:H layers are prepared at T = 200 °C, p = 0.8 Torr, and P = 0.04 W/cm^2 as a function of R varied from 10 to 50. The growth evolution diagram for intrinsic Si:H as a function of variable hydrogen dilution, $10 \leqslant R \leqslant 50$, onto *n*-layer coated BRs has been developed and is shown in Figure 6. The hydrogen dilution and thickness of *n*-layer was fixed at R = 50 and ~200 Å, based on protocrystallinity observed in the *n*-layer growth evolution diagram. For the intrinsic layer, R = 15 is the lowest hydrogen dilution ratio at which the a→(a+nc) transition is observed within ~3000 Å of layer growth. The decrease in the (a+nc)→nc thickness with R may indicate higher nucleation density of crystallites for higher hydrogen dilution. Hence, R = 10 is identified here as optimized for *n-i-p* a-Si:H solar cells incorporating a ~3000 Å thick protocrystalline absorber [2].

The thickness of the *p*-layer should be thin enough to maximize transparency but thick enough to generate an electric field in the intrinsic layer. Typical *p*-layer thicknesses are ~100–150 Å, and a large optical band gap assists in minimizing parasitic absorption of incident light within this layer. Within the amorphous and protocrystalline phase the band gap of the *p*-layer generally increases with increasing R. The intrinsic layer, *p*-layer, and their interface are most directly responsible for open circuit voltage optimization, which can be guided using growth evolution diagrams [63,64]. The *p*-type Si:H layers are prepared at T = 100°C, p = 1.5 Torr, P = 0.066 W/cm^2, and D = 0.0125 as a function of R varied from 50 to 200 on borosilicate glass initially coated with ~3000 Å thick intrinsic a-Si:H prepared at R = 10. From the growth evolution diagram, it can be observed that the *p*-layer depositions with R > 150 nucleate crystallites within the typical *p*-layer thickness used in a-Si:H based PV. The R = 110 film grows initially as a-Si:H and the a→(a+nc) transition occurs after a bulk layer thickness of 545 Å. Depositions at $50 \leqslant R \leqslant$ 100 indicate that this transition occurs for thicknesses greater than the deposited 650 Å, which is outside the range of interest for solar cells. At R = 200, the a→(a+nc) transition occurs at a bulk thickness of 40 Å, and the (a+nc)→nc transition occurs within 200 Å. The *p*-layer should be deposited at the maximum R that can be sustained without crossing the a→(a+nc) transition boundary throughout the desired thickness of 100–150 Å here. This *p*-layer growth evolution diagram is comparable to previously published diagrams [11,12,63,64].

The slope of d_b, r(t) = d(d_b(t))/dt, was used to determine the deposition rate of each film even though ε for films containing nanocrystallites are not accurate due to phase evolution with thickness. Figure 7 shows variations in growth rate as functions of R for *n-*, *i-*, and *p*-layers. The deposition rate shows a familiar trend in that it decreases with increasing R. Increased atomic hydrogen present in the plasma resulting from the increase in hydrogen dilution may etch weakly bonded material, leading to the removal of potentially defect-rich material and slowing the deposition rate. These deposition rates were later used in VIA of RTSE data collected for films nucleating crystallites. A schematic diagram showing a single junction n-i-p device with R optimized for the intended thicknesses of each a-Si:H layer is shown in Figure 8.

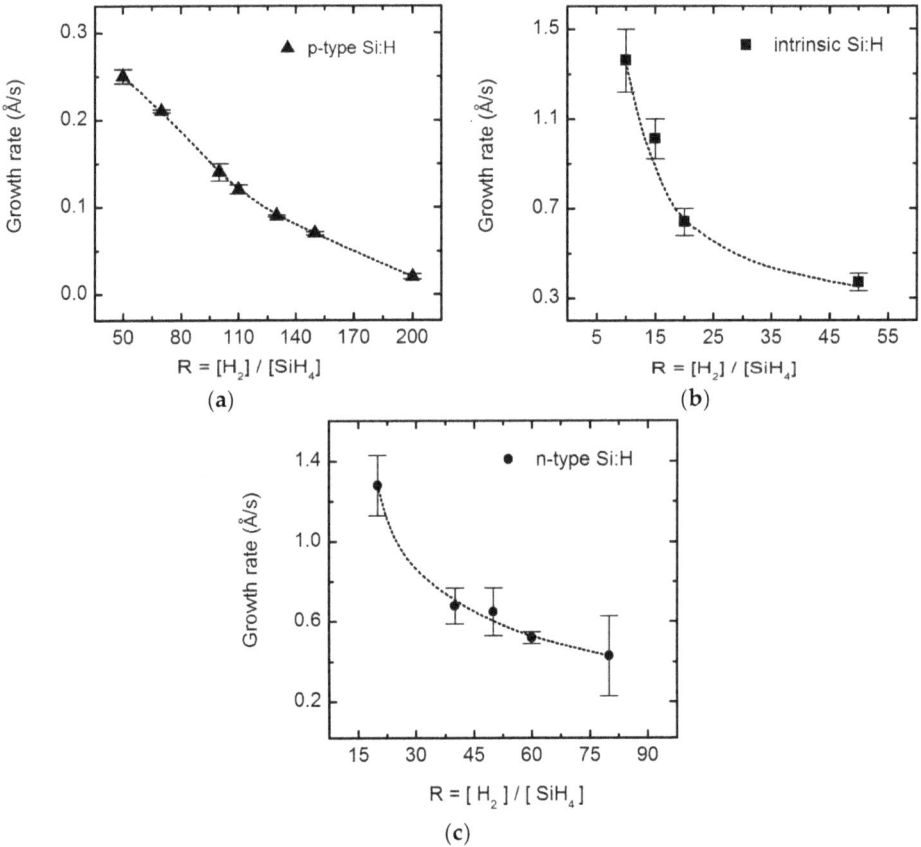

Figure 7. Deposition rates of (a) n-, (b) i-, and (c) p-layers on ZnO/Ag, n-layer/ZnO/Ag, and i-layer/glass, respectively, as functions of R.

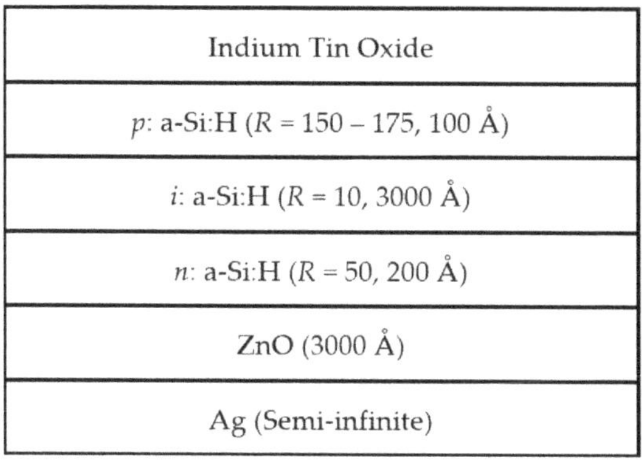

Indium Tin Oxide
p: a-Si:H (R = 150 – 175, 100 Å)
i: a-Si:H (R = 10, 3000 Å)
n: a-Si:H (R = 50, 200 Å)
ZnO (3000 Å)
Ag (Semi-infinite)

Figure 8. Schematic of a single junction a-Si:H based solar cell prepared in the *n-i-p* configuration. Each amorphous or protocrystalline Si:H layer is optimized to a value of *R* with an intended thickness.

3.3. Ex Situ SE Study of a-Si:H in n-i-p Configuration Solar Cells from the Mid-IR to Near UV

Ellipsometric spectra from 0.04 to 5.0 eV were collected and analyzed for a ZnO/Ag BR over-coated with intrinsic a-Si:H and *n*-type a-Si:H layers. This ZnO/Ag BR sample was over-coated with a-Si:H to determine ε for a-Si:H over the 0.04 to 5.0 eV range as well as identify modifications to the underlying ZnO due to this over-deposition. The structural model for the a-Si:H coated ZnO/Ag BR consisted of a semi-infinite opaque Ag metal layer, a 108 Å ZnO + Ag interfacial layer with fixed thickness from the previous analysis given in Table 3, an average 2751 $\pm$ 5 Å bulk ZnO layer produced by the mean d_b values obtained from the simultaneous fitting of the IR and the near IR-UV range spectra, a 84 $\pm$ 1 Å 0.5 *n*-type a-Si:H + 0.5 ZnO Bruggeman effective medium approximation interfacial layer, a 278 $\pm$ 1 Å a-Si:H *n*-layer, a 30 $\pm$ 1 Å 0.5 intrinsic + 0.5 *n*-type a-Si:H Bruggeman effective medium approximation interfacial layer, a 3621 $\pm$ 2 Å bulk intrinsic a-Si:H layer, and a 29 $\pm$ 1 Å surface roughness represented using Bruggeman effective medium approximation of 0.5 intrinsic a-Si:H/0.5 void volume fractions. The *n*-layer + ZnO interface, *n*-layer bulk layer, and *n*-layer surface roughness thicknesses are obtained from *in situ* RTSE measurements and analysis prior to intrinsic a-Si:H deposition. The intrinsic + *n*-type a-Si:H interface thickness is set at the same value as the *n*-layer surface roughness assuming that over-deposited intrinsic a-Si:H fill the voids in the *n*-layer surface. Parameters describing ε for ZnO and a-Si:H are listed in Table 5. As with the IR extended analysis of the ZnO/Ag sample, a common parameterization of ε for the materials over the full spectral range was applied, the bulk layer thicknesses for the ZnO and intrinsic a-Si:H layers were fit independently for spectra collected from each instrument, and all other layer thicknesses were either fixed from prior analyses or kept common between the two sets of spectra. For the i-layer, the nominal substrate temperature and hydrogen dilution ratio were $T = 200°C$ and $R = 10$, respectively. The optimized n-type a-Si:H thickness was fixed at 278 Å for $R = 50$ as found by RTSE growth evolution studies. The depositing material fills the void space in the surface roughness layer of the underlying film. The protrusions in the surface roughness of the substrate film are coated with the depositing material, generating an interface layer associated with the growing film. Bruggeman effective medium approximation defines spectra in ε for these interfaces.

Table 5. Parameters describing ε and structure for a ZnO/Ag BR coated with n-type and intrinsic a-Si:H. Experimental ellipsometric spectra were collected *ex situ* using near infrared to ultraviolet (0.734 to 5.0 eV) and infrared (0.04 to 0.734 eV) spectral range instruments and fit jointly using least squares regression analysis with an unweighted estimator error function, $\sigma = 11 \times 10^{-3}$. Parameters describing ε for Ag and the ZnO + Ag interface were fixed from Tables 2 and 3 respectively. Parameters describing ε for the n-layer were determined from RTSE analysis of data collected at $T = 200\,^{\circ}$C, parameterized by a Cody-Lorentz oscillator, and then parameter values extrapolated to room temperature. The ZnO and intrinsic a-Si:H bulk layer thicknesses were allowed to vary separately for each set of spectra; all other parameters are common to both analyses. For ZnO, the parameterization of ε consisted of two CPPB oscillators with all parameters except the amplitudes fixed to the values in Table 4, four Lorentz oscillators, and ε_∞. For a-Si:H layers, the parameterization of ε was based on a Cody-Lorentz oscillator and ε_∞. A Sellmeier oscillator and three Gaussian oscillators were added to the parameterization of ε for intrinsic a-Si:H.

Layer	Oscillators			
	i-type a-Si:H	Cody-Lorentz E_g (T&R) = 1.780 $\pm$ 0.001; ε_∞ = 1.50 $\pm$ 0.01		
i-type a-Si:H		Gaussian		
d_b (Near IR to UV) = 3623 $\pm$ 1 Å	A (Unitless)	Γ (eV)	E_0 (eV)	
d_b (IR) = 3619 $\pm$ 2 Å	1.732 $\pm$ 0.06	0.013 $\pm$ 0.001	0.079 $\pm$ 0.001	
d_s = 29.0 $\pm$ 0.3 Å	0.28 $\pm$ 0.01	0.010 $\pm$ 0.001	0.249 $\pm$ 0.001	
	0.41 $\pm$ 0.04	0.016 $\pm$ 0.002	0.106 $\pm$ 0.001	
		Sellmeier		
	0.0050 $\pm$ 0.0002 eV^2	-	0	

Layer	Oscillators				
i- a-Si:H/*n*-type a-Si:H	*n*-type a-Si:H	Cody-Lorentz; ε_∞ = 1			
Interface = 30 $\pm$ 1 Å					
n-layer	A (eV)	Γ(eV)	E_0 (eV)	E_g (eV)	E_p (eV)
d_b = 278 $\pm$ 1 Å					
n-type a-Si:H/ZnO					
Interface = 84 Å	62	2.01	3.99	1.65	1.05

Layer	Oscillators			
	ZnO	CPPB (μ = 0.5) ε_∞ = 1.91 $\pm$ 0.02		
	A	Γ (eV)	E_n (eV)	Θ
ZnO	4.04 $\pm$ 0.05	0.209	3.364	-20.8
d_b (Near IR to UV) = 2763 $\pm$ 3 Å	1.31 $\pm$ 0.02	3.95	3.94	0
d_b (IR) = 2738 $\pm$ 5 Å		Lorentz		
	A (Unitless)	Γ (eV)	E_0 (eV)	
	3.89 $\pm$ 0.1	0.233 $\pm$ 0.001	0.162 $\pm$ 0.002	
	82.0 $\pm$ 4.0	0.0030 $\pm$ 0.0003	0.0506 $\pm$ 0.0001	
	16.4 $\pm$ 0.4	0.039 $\pm$ 0.002	0.085 $\pm$ 0.001	
	13.0 $\pm$ 3.0	0.004 $\pm$ 0.002	0.047 $\pm$ 0.001	

3.3.1. Phonon Mode Variations in ZnO

Differences in ε for the ZnO are expected when over-coated with a-Si:H. PECVD of a-Si:H raises the temperature of ZnO to 200°C and exposes it to hydrogen in the plasma. There are many studies on the growth and various effects of annealing on the optical and structural properties of ZnO layers [65–70]. It is well known that the properties of ZnO layers are strongly affected by not only the deposition conditions but also the post-deposition annealing conditions or temperature treatments. Annealing has a large effect on the crystallinity of the layers in terms of grain size, residual strain, and the defect density as compared to as-deposited films. As noted in Table 5, the amplitude of the CPPB oscillators of ZnO over-coated with a-Si:H are fit to account for changes in ε occurring during PECVD of the a-Si:H layers. The increase in amplitude for higher energy absorption features in ε and the decrease in film thickness compared to the sample without a-Si:H coating at $T = 200$ °C indicate that the as-deposited ZnO film densifies and increases the degree of crystallinity after annealing at the a-Si:H deposition temperature. These variations are generally consistent with literature [69,70], in that the imaginary component of the optical response related to electronic transitions increases in amplitude and sharpens. However, Liu *et al.* reports a decrease in the real part of the complex index of refraction in the transparent region, which they attribute to void forming along with larger crystalline grains. In our samples, void formation is not observed optically, however the decrease in thickness implies that crystallite growth has occurred which coupled with the higher observed real part of ε indicates that this film is now overall more densely packed. The comparison of different phonon modes in ZnO with and without a-Si:H coating is shown in Figure 9. The characteristic TO modes with A_1 and E_1 symmetry at 0.0467 eV (376.66 cm^{-1}) and 0.0506 eV (408.12 cm^{-1}) are present and able to be resolved [51–53]. The vibrational mode at 0.0847 eV (683.15 cm^{-1}) can be attributed to longitudinal optical (LO) mode with E_1 symmetry [52]. The orientation of grains in the film could be a reason for shifting of modes to slightly higher or lower wavenumbers. The splitting of the peak observed at 404 cm^{-1} for the uncoated ZnO into two expected peaks at 377 and 408 cm^{-1} for ZnO over-coated with a-Si:H and appearance of the 683 cm^{-1} mode is likely due to grain size increases or a reduction in defect density from annealing at the a-Si:H deposition temperature of 200°C [68]. The increase in amplitude for ε_2 of phonon mode at 408 cm^{-1} also supports the idea that grain restructuring and material densification occurs. The presence of an additional absorption mode at 1306.62 cm^{-1} (0.162 eV) can be associated with oxygen-hydrogen (O-H) bonds in the thin film, such as the formation of zinc hydroxide or absorbed water or stretching modes of hydrogen bonded to heavier elements like zinc [54]. The large broadening of this absorption peak could be due to the modification or damage to the ZnO as a result of exposure to hydrogen in the plasma.

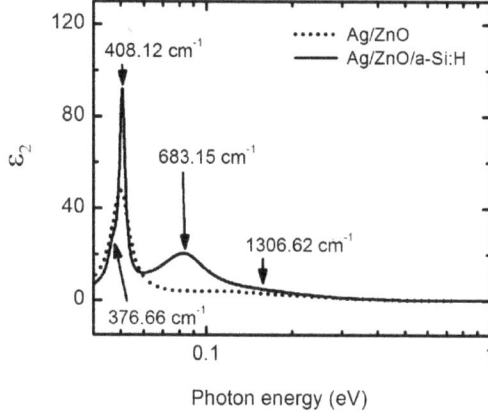

Figure 9. Comparison of lower energy features in ε_2 as a function of photon energy for ZnO with (solid line) and without (dotted line) over-deposition of a-Si:H. Parameters describing the sample without and with over-deposition of are listed in Tables 4 and 5 respectively.

3.3.2. Chemical Bonding in a-Si:H

After ZnO deposition, a 278 Å thick n-layer was deposited onto a ZnO/Ag coated substrate with deposition conditions given in Table 1. The n-layer optical properties, as well as its d_b and d_s, were obtained from RTSE analysis. The final numerically inverted spectra in ε for the n-layer were fit to a Cody-Lorentz oscillator [71]. The Cody-Lorentz oscillator is described by:

$$\varepsilon_2(E) = \begin{cases} \dfrac{AE_0\Gamma E}{\left(E^2 - E_0^2\right)^2 + \Gamma^2 E^2} \dfrac{\left(E - E_g\right)^2}{\left(E - E_g\right)^2 + E_p^2} & E > E_g \\ 0 & E \leqslant E_g \end{cases}, \qquad (7)$$

and

$$\varepsilon_1(E) = \frac{2}{\pi} P \int_0^\infty \frac{\xi\,\varepsilon_2(\xi)}{\xi^2 - E^2} d\xi \qquad (8)$$

where A is the amplitude, Γ is the broadening, E_0 is the resonance energy, E_g represents an absorption onset determined from a parabolic band constant dipole matrix element, and $E_p + E_g$ represents the transition between Cody gap-like and Lorentz-like behavior. Analytical Kramers-Kronig transformation of ε_2 yields ε_1. Parameters describing ε for the n-layer at the deposition temperature $T = 200\ °C$ are $A = 59 \pm 2$ eV, $\Gamma = 2.12 \pm 0.02$ eV, $E_0 = 3.99 \pm 0.01$ eV, $E_g = 1.58 \pm 0.04$ eV, and $E_p = 0.96 \pm 0.09$ eV.

87

Figure 10 shows spectra in ε for the $R = 10$ a-Si:H intrinsic layer parameterized using a Cody-Lorentz oscillator at high energies and Gaussian oscillators to represent the IR vibrational modes. Each Gaussian oscillator [72] is described by:

$$\varepsilon_2(E) = Ae^{-(\frac{E - E_n}{\sigma})^2} - Ae^{-(\frac{E + E_n}{\sigma})^2} \tag{9}$$

$$\sigma = \frac{\Gamma}{2\sqrt{\ln(2)}} \tag{10}$$

where A, Γ, and E_n represent amplitude, broadening, and resonance energy respectively, and ε_1 is generated by Kramers-Kronig transformation of ε_2 (Equation 8). Fit parameters are listed in Table 5. The Cody-Lorentz oscillator parameters for intrinsic a-Si:H were linked to a single fit parameter, E_g from transmission and reflection spectroscopy, by linear relationships previously determined for PV device quality a-Si:H [71]. This technique minimizes the number of fit parameters allowing for extraction of physically realistic ε. Parameters describing spectra in ε for the underlying n-layer were extrapolated based on previously observed trends in the Cody-Lorentz oscillator parameters with temperature [73].

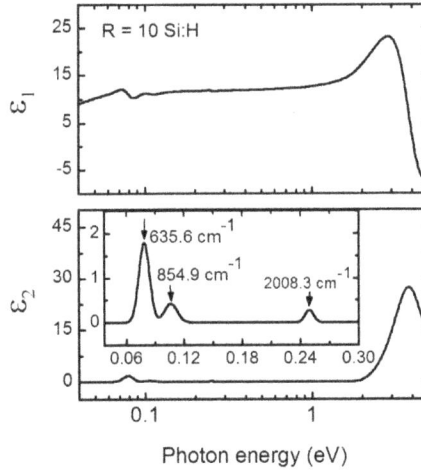

Figure 10. Spectra in ε (top panel, real part ε_1; bottom panel, imaginary part ε_2) extracted over a spectral range from 0.04 to 5 eV for 3621 ± 2 Å $R = 10$ a-Si:H films on BR over-coated with a $R = 50$ n-layer. The inset shows lower energy features in ε_2 as a function of photon energy representing Si-H$_n$ vibrational modes as modeled by Gaussian oscillators.

IR vibrational studies of a-Si:H have been useful in understanding the role of Si-H bonding in determining a-Si:H properties. High mobility and reactivity of hydrogen enables passivation of the electronic defect states in a-Si:H and relaxes the a-Si:H network to improve electronic and structural properties. IR-absorption studies have shown that hydrogen in a-Si:H is bonded as Si-H$_n$, with n = 1, 2, and 3 [35,74]. IR features in ε for the intrinsic a-Si:H film are highlighted in the inset of Figure 10. Spectra in ε for a-Si:H in the n-i-p device configuration exhibited bending modes near 0.079 eV (635.6 cm^{-1}) and a stretching monohydride (Si-H) mode around 0.249 eV (2008.3 cm^{-1}). In addition to the expected Si-H modes, this a-Si:H sample exhibited an absorption mode centered around 0.106 eV (854.9 cm^{-1}), which can be attributed to the bending or scissoring Si-H$_2$ dihydride mode. The peak centered ~2100 cm^{-1} assigned to the dihydride (Si-H$_2$) or clustered hydrogen is not observed. Although Si-H and Si-H$_2$ bonding modes were previously resolved in ellipsometric measurements for other samples [75], we can resolve only the Si-H peak here possibly due to a much lower amplitude of the Si-H$_2$ peak or reduced sensitivity to that feature in this particular sample. In addition to mode deconvolution in ε, the Si-H$_2$ mode is also not observed in the extinction coefficient, k, or the absorption coefficient, α, obtained from ε. The absence of that peak usually confirms the presence of ordered dense Si:H material [76,77]. The amplitude of ε_1 and the relatively high amplitude of the near IR to UV absorption feature in ε_2 indicate that this is dense material and suitable for PV devices.

4. Summary and Conclusions

RTSE and IR-SE have been demonstrated as a useful metrology technique for characterization of PECVD Si:H layers and components of the BR structure used in n-i-p a-Si:H solar cells. Growth evolution diagrams were developed for n-type, intrinsic, and p-type Si:H to identify the regions of optimized protocrystalline a-Si:H material for the respective thicknesses used in the solar cell configuration. IR to UV *ex situ* SE measurements and analysis were used to determine spectra in ε for Ag, ZnO, the ZnO + Ag interface, and protocrystalline intrinsic a-Si:H in the device configuration. Free carrier absorption in Ag and the ZnO + Ag interface, the particle plasmon feature in the ZnO/Ag interface, and four IR phonon modes in ZnO were identified. Si-H$_n$ bonding modes were identified in ε obtained from intrinsic a-Si:H prepared on a n-type a-Si:H coated BR. IR-SE has been demonstrated to be sensitive to bonding characteristics of a-Si:H layers in the PV device configuration. Overall, the results and analysis procedures developed here are applicable to more directly relating film properties, as obtained by non-destructive measurements in the PV device configuration, with variations in device performance.

Acknowledgments: We gratefully acknowledge support from the University of Toledo start up funds and the Ohio Department of Development (ODOD) Ohio Research Scholar Program entitled Northwest Ohio Innovators in Thin Film Photovoltaics, Grant No. TECH 09-025.

Author Contributions: Laxmi Karki Gautam primarily fabricated samples, collected and analyzed RTSE data, and analyzed *ex situ* SE data. Maxwell M. Junda fabricated samples and performed RTSE data collection and analysis. Hamna F. Haneef performed IR-SE measurements and data analysis. Robert W. Collins and Nikolas J. Podraza guided the experiment design and data analysis. Laxmi Karki Gautam and Nikolas J. Podraza wrote the paper with input from Maxwell M. Junda, Hamna F. Haneef, and Robert W. Collins.

Conflicts of Interest: The authors have no conflict of interest. The funding sponsors had no role in the design of the study, in the collection, analyses, or interpretation of data; in the writing of the manuscript; and in the decision to publish the results.

References

1. Robertson, J. Deposition mechanism of hydrogenated amorphous silicon. *J. Appl. Phys.* **2000**, *87*, 2608–2617.

2. Collins, R.; Ferlauto, A.; Ferreira, G.; Chen, C.; Koh, J.; Koval, R.; Lee, Y.; Pearce, J.; Wronski, C. Evolution of microstructure and phase in amorphous, protocrystalline, and microcrystalline silicon studied by real time spectroscopic ellipsometry. *Sol. Energy Mater. Sol. Cells* **2003**, *78*, 143–180.

3. Robertson, J. Growth mechanism of hydrogenated amorphous silicon. *J. Non-Cryst. Solids* **2000**, *266*, 79–83.

4. Fujiwara, H.; Kondo, M.; Matsuda, A. Real-time spectroscopic ellipsometry studies of the nucleation and grain growth processes in microcrystalline silicon thin films. *Phys. Rev. B* **2001**, *63*.

5. Schiff, E.A.; Deng, X. Amorphous silicon-based solar cells. In *Handbook of Photovoltaic Science and Engineering*; Luque, S.H.A., Ed.; Wiley: New York, NY, USA, 2003; pp. 487–545.

6. Guha, S.; Cohen, D.; Schiff, E.; Stradins, P.; Taylor, P.; Yang, J. Industry-academia partnership helps drive commercialization of new thin-film silicon technology. *Photovolt. Int.* **2011**, *134*. Available online: http://citeseerx.ist.psu.edu/viewdoc/download?doi=10.1.1.397.3368&rep=rep1&type=pdf (accessed on 1 September 2011).

7. Koh, J.; Ferlauto, A.; Rovira, P.; Wronski, C.; Collins, R. Evolutionary phase diagrams for plasma-enhanced chemical vapor deposition of silicon thin films from hydrogen-diluted silane. *Appl. Phys. Lett.* **1999**, *75*, 2286–2288.

8. Matsui, T.; Sai, H.; Saito, K.; Kondo, M. High-efficiency thin-film silicon solar cells with improved light-soaking stability. *Prog. Photovolt. Res. Appl.* **2013**, *21*, 1363–1369.

9. Sai, H.; Matsui, T.; Matsubara, K.; Kondo, M.; Yoshida, I. 11.0%-Efficient thin-film microcrystalline silicon solar cells with honeycomb textured substrates. *IEEE J. Photovolt.* **2014**, *4*, 1349–1353.

10. Yan, B.; Yue, G.; Sivec, L.; Yang, J.; Guha, S.; Jiang, C.-S. Innovative dual function nc-SiO$_x$:H layer leading to a >16% efficient multi-junction thin-film silicon solar cell. *Appl. Phys. Lett.* **2011**, *99*.

11. Rovira, P.; Ferlauto, A.; Koval, R.; Wronski, C.; Collins, R.; Ganguly, G. Real time optics of p-type silicon deposition on specular and textured ZnO surfaces. In Proceedings of the Conference Record of the Twenty-Eighth IEEE Photovoltaic Specialists Conference, Anchorage, AK, USA, 15–22 September 2000; pp. 772–775.

12. Koval, R.; Chen, C.; Ferreira, G.; Ferlauto, A.; Pearce, J.; Rovira, P.; Wronski, C.; Collins, R. Maximization of the open circuit voltage for hydrogenated amorphous silicon nip solar cells by incorporation of protocrystalline silicon p-type layers. *Appl. Phys. Lett.* **2002**, *81*, 1258–1260.

13. Ferreira, G.; Ferlauto, A.; Pearce, J.; Wronski, C.; Ross, C.; Collins, R. Comparison of phase diagrams for vhf and rf plasma-enhanced chemical vapor deposition of Si: H films. In *MRS Proceedings*; Cambridge University Press: Cambridge, UK, 2004; pp. A5.2.1–A5.2.6.

14. Stoke, J.; Dahal, L.; Li, J.; Podraza, N.; Cao, X.; Deng, X.; Collins, R. Optimization of Si:H multijunction nip solar cells through the development of deposition phase diagrams. In Proceedings of the 33rd IEEE Photovoltaic Specialists Conference, PVSC '08, San Diego, CA, USA, 11–16 May 2008; pp. 1–6.

15. Dahal, L.R.; Huang, Z.; Attygalle, D.; Sestak, M.N.; Salupo, C.; Marsillac, Y.; Collins, R. Application of real time spectroscopic ellipsometry for analysis of roll-to-roll fabrication of Si:H solar cells on polymer substrates. In Proceedings of the 2010 35th IEEE Photovoltaic Specialists Conference (PVSC), Honolulu, HI, USA, 20–25 June 2010; pp. 000631–000636.

16. Huang, Z.; Dahal, L.; Salupo, C.; Ferlauto, A.; Podraza, N.J.; Collins, R.W. Optimization of a-Si:H pin solar cells through development of n-layer growth evolution diagram and large area mapping. In Proceedings of the 2013 IEEE 39th Photovoltaic Specialists Conference (PVSC), Tampa, FL, USA, 16–21 June 2013; pp. 1788–1793.

17. Murata, D.; Yuguchi, T.; Fujiwara, H. Characterization of μc-Si:H/a-Si:H tandem solar cell structures by spectroscopic ellipsometry. *Thin Solid Films* **2014**, *571*, 756–761.

18. Junda, M.M.; Shan, A.; Koirala, P.; Collins, R.W.; Podraza, N.J. Spectroscopic ellipsometry applied in the full pin a-Si:H solar cell device configuration. *IEEE J. Photovolt.* **2015**, *5*, 307–312.

19. Yue, G.; Yan, B.; Ganguly, G.; Yang, J.; Guha, S.; Teplin, C.W. Material structure and metastability of hydrogenated nanocrystalline silicon solar cells. *Appl. Phys. Lett.* **2006**, *88*.

20. Yue, G.; Sivec, L.; Owens, J.M.; Yan, B.; Yang, J.; Guha, S. Optimization of back reflector for high efficiency hydrogenated nanocrystalline silicon solar cells. *Appl. Phys. Lett.* **2009**, *95*.

21. Vetterl, O.; Finger, F.; Carius, R.; Hapke, P.; Houben, L.; Kluth, O.; Lambertz, A.; Mück, A.; Rech, B.; Wagner, H. Intrinsic microcrystalline silicon: A new material for photovoltaics. *Sol. Energy Mater. Sol. Cells* **2000**, *62*, 97–108.

22. Shah, A.; Meier, J.; Vallat-Sauvain, E.; Wyrsch, N.; Kroll, U.; Droz, C.; Graf, U. Material and solar cell research in microcrystalline silicon. *Sol. Energy Mater. Sol. Cells* **2003**, *78*, 469–491.

23. Dahal, L.R.; Li, J.; Stoke, J.A.; Huang, Z.; Shan, A.; Ferlauto, A.S.; Wronski, C.R.; Collins, R.W.; Podraza, N.J. Applications of real-time and mapping spectroscopic ellipsometry for process development and optimization in hydrogenated silicon thin-film photovoltaics technology. *Sol. Energy Mater. Sol. Cells* **2014**, *129*, 32–56.

24. Shan, A.; Fried, M.; Juhasz, G.; Major, C.; Polgár, O.; Németh, Á.; Petrik, P.; Dahal, L.R.; Chen, J.; Huang, Z. High-speed imaging/mapping spectroscopic ellipsometry for in-line analysis of roll-to-roll thin-film photovoltaics. *IEEE J. Photovolt.* **2014**, *4*, 355–361.

25. Aryal, P.; Pradhan, P.; Attygalle, D.; Ibdah, A.-R.; Aryal, K.; Ranjan, V.; Marsillac, S.; Podraza, N.J.; Collins, R.W. Real-time, in-line, and mapping spectroscopic ellipsometry for applications in Cu (in Ga) Se metrology. *IEEE J. Photovolt.* **2014**, *4*, 333–339.

26. Koirala, P.; Tan, X.; Li, J.; Podraza, N.J.; Marsillac, S.; Rockett, A.; Collins, R.W. Mapping spectroscopic ellipsometry of CdTe solar cells for property-performance correlations. In Proceedings of the 2014 IEEE 40th Photovoltaic Specialist Conference (PVSC), Denver, CO, USA, 8–13 June 2014; pp. 0674–0679.

27. Dahal, L.R.; Sainju, D.; Li, J.; Stoke, J.A.; Podraza, N.; Deng, X.; Collins, R.W. Plasmonic characteristics of Ag/ZnO back-reflectors for thin film Si photovoltaics. In Proceedings of the 33rd IEEE Photovoltaic Specialists Conference, PVSC '08, San Diego, CA, USA, 11–16 May 2008; pp. 1–6.

28. Dahal, L.R.; Sainju, D.; Podraza, N.; Marsillac, S.; Collins, R. Real time spectroscopic ellipsometry of Ag/ZnO and Al/ZnO interfaces for back-reflectors in thin film Si:H photovoltaics. *Thin Solid Films* **2011**, *519*, 2682–2687.

29. Oldham, W. Numerical techniques for the analysis of lossy films. *Surf. Sci.* **1969**, *16*, 97–103.

30. Aspnes, D. Minimal-data approaches for determining outer-layer dielectric responses of films from kinetic reflectometric and ellipsometric measurements. *Appl. Phys. Lett.* **1993**, *62*, 343–345.

31. Ferlauto, A.; Ferreira, G.; Koval, R.; Pearce, J.; Wronski, C.; Collins, R.; Al-Jassim, M.; Jones, K. Evaluation of compositional depth profiles in mixed-phase (amorphous + crystalline) silicon films from real time spectroscopic ellipsometry. *Thin Solid Films* **2004**, *455*, 665–669.

32. Karki Gautam, L.; Haneef, H.; Junda, M.; Saint John, D.; Podraza, N. Approach for extracting complex dielectric function spectra in weakly-absorbing regions. *Thin Solid Films* **2014**, *571*, 548–553.

33. Podraza, N.J.; Saint John, D.B.; Ko, S.W.; Schulze, H.M.; Li, J.; Dickey, E.C.; Trolier-McKinstry, S. Optical and structural properties of solution deposited nickel manganite thin films. *Thin Solid Films* **2011**, *519*, 2919–2923.

34. Knights, J.C.; Lucovsky, G.; Nemanich, R.J. Defects in plasma-deposited a-Si:H. *J. Non-Cryst. Solids* **1979**, *32*, 393–403.

35. Langford, A.A.; Fleet, M.L.; Nelson, B.P.; Lanford, W.A.; Maley, N. Infrared absorption strength and hydrogen content of hydrogenated amorphous silicon. *Phys. Rev. B* **1992**, *45*, 13367–13377.

36. Podraza, N.J.; Saint John, D.B. Optical characterization of structurally graded $Si_{1-x}Ge_x$:H thin films. In Proceedings of the 2012 38th IEEE Photovoltaic Specialists Conference (PVSC), Austin, TX, USA, 3–8 June 2012; pp. 000354–000359.

37. Dahal, L.R.; Zhiquan, H.; Attygalle, D.; Salupo, C.; Marsillac, S.; Podraza, N.J.; Collins, R.W. Correlations between mapping spectroscopic ellipsometry results and solar cell performance for evaluations of nonuniformity in thin-film silicon photovoltaics. *IEEE J. Photovolt.* **2013**, *3*, 387–393.

38. Lee, J.; Rovira, P.; An, I.; Collins, R. Rotating-compensator multichannel ellipsometry: Applications for real time Stokes vector spectroscopy of thin film growth. *Rev. Sci. Instrum.* **1998**, *69*, 1800–1810.

39. Johs, B.D.; Woollam, J.A.; Herzinger, C.M.; Hilfiker, J.N.; Synowicki, R.A.; Bungay, C.L. Overview of variable-angle spectroscopic ellipsometry (VASE): II. In *Advanced Applications, Society of Photo-Optical Instrumentation Engineers (SPIE) Conference Series*; Society of Photo-Optical Instrumentation Engineers: Bellingham, WA, USA, 1999; pp. 29–58.

40. Johs, B.; Herzinger, C.M. Quantifying the accuracy of ellipsometer systems. *Phys. Status Solidi* **2008**, *5*, 1031–1035.

41. Hilfiker, J.N.; Bungay, C.L.; Synowicki, R.A.; Tiwald, T.E.; Herzinger, C.M.; Johs, B.; Pribil, G.K.; Woollam, J.A. Progress in spectroscopic ellipsometry: Applications from vacuum ultraviolet to infrared. *J. Vac. Sci. Technol. A* **2003**, *21*, 1103–1108.

42. Aspnes, D.E. Optical properties of thin films. *Thin Solid Films* **1982**, *89*, 249–262.

43. Fujiwara, H.; Koh, J.; Rovira, P.; Collins, R. Assessment of effective-medium theories in the analysis of nucleation and microscopic surface roughness evolution for semiconductor thin films. *Phys. Rev. B* **2000**, *61*.

44. Azzam, R.; Bashara, N. *Ellipsometry and Polarized Light*; North-Holland: Amsterdam, The Netherlands, 1981.

45. Tiwald, T.E.; Thompson, D.W.; Woollam, J.A.; Paulson, W.; Hance, R. Application of IR variable angle spectroscopic ellipsometry to the determination of free carrier concentration depth profiles. *Thin Solid Films* **1998**, *313–314*, 661–666.

46. Aspnes, D.E. *Handbook on Semiconductors*; Balkanski, A.M., Ed.; North-Holland: Amsterdam, The Netherlands, 1980; pp. 125–127.

47. Collins, R.W.; Ferlauto, A.S. *Handbook of Ellipsometry*; Tompkins, H., Irene, E.A., Eds.; William Andrew: Norwich, NY, USA, 2005; pp. 159–171.

48. Sainju, D.; van den Oever, P.J.; Podraza, N.J.; Syed, M.; Stoke, J.A.; Jie, C.; Xiesen, Y.; Xunming, D.; Collins, R.W. Origin of optical losses in Ag/ZnO back-reflectors for thin film Si photovoltaics. In Proceedings of the Conference Record of the 2006 IEEE 4th World Conference on Photovoltaic Energy Conversion, Waikoloa, HI, USA, 7–12 May 2006; pp. 1732–1735.

49. Ferlauto, A.S.; Collins, R.W. *Handbook of Ellipsometry*; Tompkins, H., Irene, E.A., Eds.; William Andrew: Norwich, NY, USA, 2005; pp. 125–129.

50. Jellison, G.E.; Boatner, L.A. Optical functions of uniaxial ZnO determined by generalized ellipsometry. *Phys. Rev. B* **1998**, *58*, 3586–3589.

51. Bundesmann, C.; Ashkenov, N.; Schubert, M.; Rahm, A.; Wenckstern, H.; Kaidashev, E.; Lorenz, M.; Grundmann, M. Infrared dielectric functions and crystal orientation of a-plane ZnO thin films on r-plane sapphire determined by generalized ellipsometry. *Thin Solid Films* **2004**, *455*, 161–166.

52. Ashkenov, N.; Mbenkum, B.N.; Bundesmann, C.; Riede, V.; Lorenz, M.; Spemann, D.; Kaidashev, E.M.; Kasic, A.; Schubert, M.; Grundmann, M.; *et al.* Infrared dielectric functions and phonon modes of high-quality ZnO films. *J. Appl. Phys.* **2003**, *93*, 126–133.

53. Tzolov, M.; Tzenov, N.; Dimova-Malinovska, D.; Kalitzova, M.; Pizzuto, C.; Vitali, G.; Zollo, G.; Ivanov, I. Vibrational properties and structure of undoped and Al-doped ZnO films deposited by RF magnetron sputtering. *Thin Solid Films* **2000**, *379*, 28–36.

54. Keyes, B.; Gedvilas, L.; Li, X.; Coutts, T. Infrared spectroscopy of polycrystalline ZnO and ZnO:N thin films. *J. Cryst. Growth* **2005**, *281*, 297–302.

55. Saint John, D.S.; Shin, H.-B.; Lee, M.-Y.; Ajmera, S.; Syllaios, A.; Dickey, E.; Jackson, T.; Podraza, N. Influence of microstructure and composition on hydrogenated silicon thin film properties for uncooled microbolometer applications. *J. Appl. Phys.* **2011**, *110*, 033714.

56. Podraza, N.; Li, J.; Wronski, C.; Dickey, E.; Collins, R. Analysis of controlled mixed-phase (amorphous + microcrystalline) silicon thin films by real time spectroscopic ellipsometry. *J. Vac. Sci. Technol. A* **2009**, *27*, 1255–1259.

57. Podraza, N.; Li, J.; Wronski, C.; Dickey, E.; Horn, M.; Collins, R. Analysis of $Si_{1-x}Ge_x$:H thin films with graded composition and structure by real time spectroscopic ellipsometry. *Phys. Status Solidi* **2008**, *205*, 892–895.

58. Aspnes, D.; Theeten, J. Spectroscopic analysis of the interface between Si and its thermally grown oxide. *J. Electrochem. Soc.* **1980**, *127*, 1359–1365.

59. Kroll, U.; Meier, J.; Shah, A.; Mikhailov, S.; Weber, J. Hydrogen in amorphous and microcrystalline silicon films prepared by hydrogen dilution. *J. Appl. Phys.* **1996**, *80*, 4971–4975.

60. Funde, A.M.; Bakr, N.A.; Kamble, D.K.; Hawaldar, R.R.; Amalnerkar, D.P.; Jadkar, S.R. Influence of hydrogen dilution on structural, electrical and optical properties of hydrogenated nanocrystalline silicon (nc-Si:H) thin films prepared by plasma enhanced chemical vapour deposition (PE-CVD). *Sol. Energy Mater. Sol. Cells* **2008**, *92*, 1217–1223.

61. Cao, X.; Stoke, J.A.; Li, J.; Podraza, N.J.; Du, W.; Yang, X.; Attygalle, D.; Liao, X.; Collins, R.W.; Deng, X. Fabrication and optimization of single-junction nc-Si:H n-i-p solar cells using Si:H phase diagram concepts developed by real time spectroscopic ellipsometry. *J. Non-Cryst. Solids* **2008**, *354*, 2397–2402.

62. Yan, B.; Yue, G.; Yang, J.; Guha, S.; Williamson, D.; Han, D.; Jiang, C.-S. Hydrogen dilution profiling for hydrogenated microcrystalline silicon solar cells. *Appl. Phys. Lett.* **2004**, *85*, 1955–1957.

63. Wronski, C.; Collins, R. Phase engineering of a-Si:H solar cells for optimized performance. *Sol. Energy* **2004**, *77*, 877–885.

64.   Pearce, J.M.; Podraza, N.; Collins, R.W.; Al-Jassim, M.M.; Jones, K.M.; Deng, J.; Wronski, C.R. Optimization of open circuit voltage in amorphous silicon solar cells with mixed-phase (amorphous + nanocrystalline) p-type contacts of low nanocrystalline content. *J. Appl. Phys.* **2007**, *101*.

65.   Jung, M.; Lee, J.; Park, S.; Kim, H.; Chang, J. Investigation of the annealing effects on the structural and optical properties of sputtered ZnO thin films. *J. Cryst. Growth* **2005**, *283*, 384–389.

66.   Kang, H.S.; Kang, J.S.; Kim, J.W.; Lee, S.Y. Annealing effect on the property of ultraviolet and green emissions of ZnO thin films. *J. Appl. Phys.* **2004**, *95*, 1246–1250.

67.   Puchert, M.; Timbrell, P.; Lamb, R. Post deposition annealing of radio frequency magnetron sputtered ZnO films. *J.Vac. Sci. Technol. A* **1996**, *14*, 2220–2230.

68.   Rolo, A.G.; de Campos, J.A.; Viseu, T.; de Lacerda-Arôso, T.; Cerqueira, M. The annealing effect on structural and optical properties of ZnO thin films produced by rf sputtering. *Superlattices Microstruct.* **2007**, *42*, 265–269.

69.   Al Asmar, R.; Ferblantier, G.; Mailly, F.; Gall-Borrut, P.; Foucaran, A. Effect of annealing on the electrical and optical properties of electron beam evaporated ZnO thin films. *Thin Solid Films* **2005**, *473*, 49–53.

70.   Liu, Y.C.; Tung, S.K.; Hsieh, J.H. Influence of annealing on optical properties and surface structure of ZnO thin films. *J. Crystal Growth* **2006**, *287*, 105–111.

71.   Ferlauto, A.; Ferreira, G.; Pearce, J.; Wronski, C.; Collins, R.; Deng, X.; Ganguly, G. Analytical model for the optical functions of amorphous semiconductors from the near-infrared to ultraviolet: Applications in thin film photovoltaics. *J. Appl. Phys.* **2002**, *92*, 2424–2436.

72.   De Meneses, D.; Malki, M.; Echegut, P. Structure and lattice dynamics of binary lead silicate glasses investigated by infrared spectroscopy. *J. Non-Cryst. Solids* **2006**, *352*, 769–776.

73.   Podraza, N.; Wronski, C.; Horn, M.; Collins, R. Dielectric functions of a-$Si_{1-x}Ge_x$:H *vs.* Ge content, temperature, and processing: Advances in optical function parameterization. *Mater. Res. Soc. Symp. Proc.* **2006**, *A10*.

74.   Brodsky, M.; Cardona, M.; Cuomo, J. Infrared and Raman spectra of the silicon-hydrogen bonds in amorphous silicon prepared by glow discharge and sputtering. *Phys. Rev. B* **1977**, *16*.

75.   Saint John, D.B.; Shen, H.; Shin, H.-B.; Jackson, T.N.; Podraza, N.J. Infrared dielectric functions of hydrogenated amorphous silicon thin films determined by spectroscopic ellipsometry. In Proceedings of the 2012 38th IEEE Photovoltaic Specialists Conference (PVSC), Austin, TX, USA, 3–8 June 2012; pp. 003112–003117.

76.   Müllerová, J.; Jurečka, S.; Šutta, P. Optical characterization of polysilicon thin films for solar applications. *Sol. Energy* **2006**, *80*, 667–674.

77.   Smets, A.; Kessels, W.; van de Sanden, M. Vacancies and voids in hydrogenated amorphous silicon. *Appl. Phys. Lett.* **2003**, *82*, 1547–1549.

Crystal Structure Formation of $CH_3NH_3PbI_{3-x}Cl_x$ Perovskite

Shiqiang Luo and Walid A. Daoud

Abstract: Inorganic-organic hydride perovskites bring the hope for fabricating low-cost and large-scale solar cells. At the beginning of the research, two open questions were raised: the hysteresis effect and the role of chloride. The presence of chloride significantly improves the crystallization and charge transfer property of the perovskite. However, though the long held debate over of the existence of chloride in the perovskite seems to have now come to a conclusion, no prior work has been carried out focusing on the role of chloride on the electronic performance and the crystallization of the perovskite. Furthermore, current reports on the crystal structure of the perovskite are rather confusing. This article analyzes the role of chloride in $CH_3NH_3PbI_{3-x}Cl_x$ on the crystal orientation and provides a new explanation about the (110)-oriented growth of $CH_3NH_3PbI_3$ and $CH_3NH_3PbI_{3-x}Cl_x$.

Reprinted from *Materials*. Cite as: Luo, S.; Daoud, W.A. Crystal Structure Formation of $CH_3NH_3PbI_{3-x}Cl_x$ Perovskite. *Materials* **2016**, *9*, 123.

1. Introduction

Since the first organic-inorganic halide perovskite solar cell was reported [1], perovskites have attracted growing interest and the power conversion efficiency (PCE) has reached 20.1% [2]. It is not very common that a photovoltaic device can experience such a rapid development. While the structure of the cells evolved from sensitized meso-structure to planar structure [3], both inorganic and organic materials can be applied as electron and hole transfer materials [4]. Furthermore, by tuning the composition of the perovskite, the band gap can be easily modified [5]. Given the numerous advantages of perovskite, a clear understanding of the crystal structure is crucial and the role of chloride in the formation of $CH_3NH_3PbI_{3-x}Cl_x$ (hereafter, we use MA short for CH_3NH_3) is one of the most pressing topics.

It has been reported that the presence of chloride in the perovskite improves the uniformity of its layer [6] and results in an increase of the carriers' diffusion length from *ca.* 100 nm to over 1 µm [7]. However, the long held debate over of the existence of chloride in the perovskite seems to have now come to a conclusion. First, when synthesizing the perovskite by the one step method with precursor solution of MACl and PbI_2 (1:1 molar ratio) in anhydrous N,N-dimethylformamide (DMF), the resulting crystal is not $MAPbI_2Cl$ but a mixture of $MAPbI_3$ and $MAPbCl_3$ [8]. This provides direct evidence that chloride (Cl^-) cannot substitute iodine (I^-) in the perovskite to form a stable crystal. Then, two contradictory results were then reported. X-ray photoelectron spectroscopy (XPS) showed that the molar ratio C:N:Pb:I:Cl of the perovskite is *ca.* 1:1:1:2:1, when prepared from a precursor of MAI:PbCl$_2$ (molar ratio 3:1) [9]. On the other hand, energy dispersive X-ray (EDX) analysis showed that no Cl$^-$ was present in the perovskite prepared from PbI$_2$ + MAI + MACl [10]. Noting that the XPS was unable to determine the existence of MAPbI$_2$Cl crystal and that EDX has its detecting limitation, more precise characterizations were needed. Later on, the simultaneous Fourier transform infrared spectroscopy analysis of the expelled gas during the decomposition of MAPbI$_{3-x}$Cl$_x$ showed the presence of Cl$^-$, angle-resolved XPS [11] and X-ray fluorescence spectroscopy (XFS) [12] not only confirmed the existence of Cl$^-$, but also showed that Cl$^-$ was located at the interface between the perovskite and the electron transport TiO$_2$ layer, and not in the perovskite structure [11,12]. Moreover, scanning transmission microscopy-energy dispersive spectroscopy (STEM-EDS) detected no trace of Cl$^-$ in the perovskite. Even though there is a strong Cl$^-$ signal, no N was observed indicting the presence of only PbCl$_2$ [13]. Thus, Cl$^-$ only appears at the interface between MAPbI$_3$ and the anode. Two more reports have further confirmed this conclusion. XPS analysis showed only weak Cl$^-$ signal after etching the surface of MAPbI$_{3-x}$Cl$_x$ by a 50 nm thickness [14]. Hard X-ray photoelectron spectroscopy and fluorescence yield X-ray absorption spectroscopy showed no Cl$^-$ at the surface of MAPbI$_{3-x}$Cl$_x$ with higher average concentration of Cl throughout the perovskite layer at the deep beneath [15]. Here, we refer to MAPbI$_{3-x}$Cl$_x$ as MAPbI$_3$ that is prepared using chloride-containing precursors. However, as the condition for depositing MAPbI$_{3-x}$Cl$_x$ differs, Cl$^-$ may still remain in the resulting perovskite layer. For instance, X-ray absorption near edge structure (XANES) results showed that x = 0.05 ± 0.03 Cl atoms per formula unit remain in the films after annealing at 95 °C for 120 min [16]. The results from photothermal induced resonance (PTIR) showed that the MAPbI$_{3-x}$Cl$_x$ film consists of a mixture of Cl-rich (x_{local} < 0.3) and Cl-poor phases after a mild annealing (60 °C, 60 min) and homogeneous Cl-poorer (x_{local} < 0.06) phase upon further annealing (110 °C) [17].

In addition, first-principles calculation results provide some good explanation. For the crystal structure, Cl$^-$ concentration was found below 3%–4% [8] and if the Cl$^-$ ions enter the crystal structure, they preferentially occupy the apical positions in the PbI$_4$X$_2$ octahedra [18]. For the electronic property, while the molecular orientations of CH$_3$NH$_3^+$ result in three times larger photocurrent response than the ferroelectric photovoltaic BiFeO$_3$, Cl$^-$ substitution at the equatorial site induces a larger response than does substitution at the apical site [19]. Results also showed that, using Cl$^-$ precursor can avoid forming the PbI defects [20]. Introducing Cl$^-$ would reduce the lattice constant which can inhibit the formation of interstitial defects [21]. As excitons may be screened by collective orientational motion of the organic cations, Cl$^-$ might hinder this motion and results in better transport properties [22]. Little difference of electronic properties was represent among orthorhombic, tetragonal and cubic phases of MAPbI$_3$ [23], however, the valance-band-maximum and conduction-band-minimum states can be mainly derived from iodine ions at some unique positions, Cl$^-$ substitution can strengthen the unique position of the ions and result in more localized charge density [24]. Thus, lower carrier recombination rate and enhanced carrier transport ensued. For the interface, the (001) and (110) surfaces tend to favor hole injection to 2,2',7,7'-tetrakis(N,N-di-p-methoxyphenylamine)-9,9'-spirobifluorene (Spiro-MeOTAD), while the (100) surface facilitates electron transfer to [6,6]-phenyl C$_{61}$-butyric acid methyl ester (PCBM) [25]. A better structural matching between adjacent rows of perovskite surface halides and TiO$_2$ under coordinated titanium may be the reason for the (110)-oriented growth of MAPbI$_{3-x}$Cl$_x$ and MAPbI$_3$ [26]. Interfacial Cl$^-$ may thus further stabilize the (110) surface and modify the interface electronic structure between MAPbI$_3$ and TiO$_2$ [26].

Despite the absence of Cl$^-$ in the perovskite, it still played an important role in the crystallization process. For instance, the morphology of MAPbI$_{3-x}$Cl$_x$ was compared with MAPbI$_3$ [27] and a model in which the Cl$^-$ rich phase modifies the morphologies of perovskite was proposed and fit well with the results from scanning electron microscopy (SEM) [27]. In addition, the transmission electron microscopy (TEM) of freeze-dried perovskite MAPbI$_{3-x}$Cl$_x$ precursor solution showed the presence of PbCl$_2$ nanoparticles [28] and this is in agreement with the dynamic light scattering (DLS) investigations of MAPbI$_{3-x}$Cl$_x$ precursor solution [29]. Thus, references [28,29] further proved the model of the heterogeneous nucleation by PbCl$_2$ nanoparticles proposed in reference [27]. However, the formation mechanism of the crystal structure remains undermined and this will be discussed in the following parts of this article.

2. Methods for Fabricating MAPbI$_{3-x}$Cl$_x$

In Section 3, we discuss the crystal structure of MAPbI$_{3-x}$Cl$_x$ according to the deposition method. As the fabrication methods were discussed in detail in reference [30], here we add a brief introduction about the preparation methods of MAPbI$_{3-x}$Cl$_x$. For the one-step deposition method, MAI:PbI$_2$/PbCl$_2$ (molar ratio 1:1 or 3:1) [31,32] were dissolved in γ-butyrolactone (GBL) or DMF, spin-coated on the substrates and annealed to form perovskite. Different annealing conditions result in different morphology of the MAPbI$_{3-x}$Cl$_x$ layer. While a rapid thermal annealing at 130 °C resulted in micron-sized perovskite grains [33], two-step annealing, such as 90 °C for 30 min then at 100 °C for 2 min [34] or 60 °C then ramping to 90 °C [35], resulted in optimal PCE on poly(3,4-ethylenedioxythiophene) poly(styrene-sulfonate) (PEDOT:PSS) substrates. A full coverage of perovskite can be achieved by rapid cooling after annealing [36]. To increase the solubility of Cl$^-$, 1,8-diiodooctane [37] or other alkyl halide additives [38] or dimethyl sulfoxide [9] can be employed. Adding poly-(vinylpyrrolidone) (PVP) can also improve the surface coverage of perovskite [39]. It is interesting to note that, for $MAPbI_{3-x}Cl_x$, a simple annealing step is enough to form a good coverage [6,40], but for $MAPbI_3$, a special step, such as multi-deposition [41], adding N-cyclohexyl-2-pyrrolidone (CHP) [42], fast deposition [43–45], or air flow during spin coating [46,47], is needed.

The sequential deposition method was mainly applied for $MAPbI_3$ perovskite. In a typical synthesis, the solution of PbI_2 in DMF was spun on a substrate as the first step then the substrate was dipped in a solution of MAI in 2-propanol (IPA) to form $MAPbI_3$ crystals as the second step [48]. For the inclusion of chloride, in the first step the $PbCl_2$ can be mixed with PbI_2 in DMF or dimethyl sulfoxide (DMSO) [49–52], and/or the second step MACl can be added [53–55]. For vapor based deposition methods, the $MAPbI_{3-x}Cl_x$ can be formed by co-evaporating MAI and $PbCl_2$ onto the substrates [56,57] or by reacting $PbCl_2$ on substrates with MAI vapor [58,59].

3. The Crystal Structure Form and Formation

3.1. Crystal Structure of MAPbI₃ Layer

The parameters and transitions of phases of bulk $MAPbI_3$ were included in references [60,61]. Here, we focus on the tetragonal and cubic phases [62]. In fact, there are no critical differences between the two phases, except a slight rotation of PbI_6 octahedra along the c-axis. The atomic structures of $MAPbI_3$ of the two phases are shown in Figure 1A,B. Thus, the tetragonal phase can be treated as a pseudocubic phase with $a^* = a/\sqrt{2}$, $c^* = c/2$ [63]. Below 54 °C, the cubic phase of $MAPbI_3$ can be transformed into the tetragonal phase [60], and the opposite transition occurs by annealing at 100 °C for 15 min [41]. In Figure 1C, the X-ray diffraction (XRD) patterns of the two phases are shown. After transformation to the tetragonal phase, the (100) and (200) peaks of cubic $MAPbI_3$ split, also new (211) and (213) peaks show up. Here, we use the peak splitting as indictor for phase transformation. Analysis of the $MAPbI_{3-x}Cl_x$ usually shows the cubic phase of $MAPbI_3$, however, with a much more preference along (100) and (200). This will be discussed in the Sections 3.2 and 3.3.

Another phase which should be noted is the amorphous phase. Pair distribution function analysis of X-ray scattering showed that after annealing at 100 °C for 30 min, the $MAPbI_3$ in meso-porous TiO_2 has about 30 atom% in medium range crystalline order and the other 70 atom% in a disordered state with a coherence length of 1.4 nm [66]. The poor crystallization of the $MAPbI_3$ in meso-porous TiO_2 was studied by high-resolution TEM [67]. Quartz crystal microbalance measurements suggest that during the sequential method only half of PbI_2 is converted to $MAPbI_3$ instantly, while the other half is involved in reversible transformation with $MAPbI_3$. Additionally, the amorphous character with a very small average crystallite size may be present after the transformation as previously discussed [68]. The amorphous phase may also present during the initially deposited MAPbI$_{3-x}$Cl$_x$, as indicated by the envelope in some XRD spectra. In reference [69], the amorphous phase MA$_5$PbCl$_4$I$_3$ was also mentioned. Moreover, both XRD and photoluminescence studies of MAPbI$_2$Cl (2MAPbI$_3$+MAPbCl$_3$) indicate the existence of the amorphous phase [70].

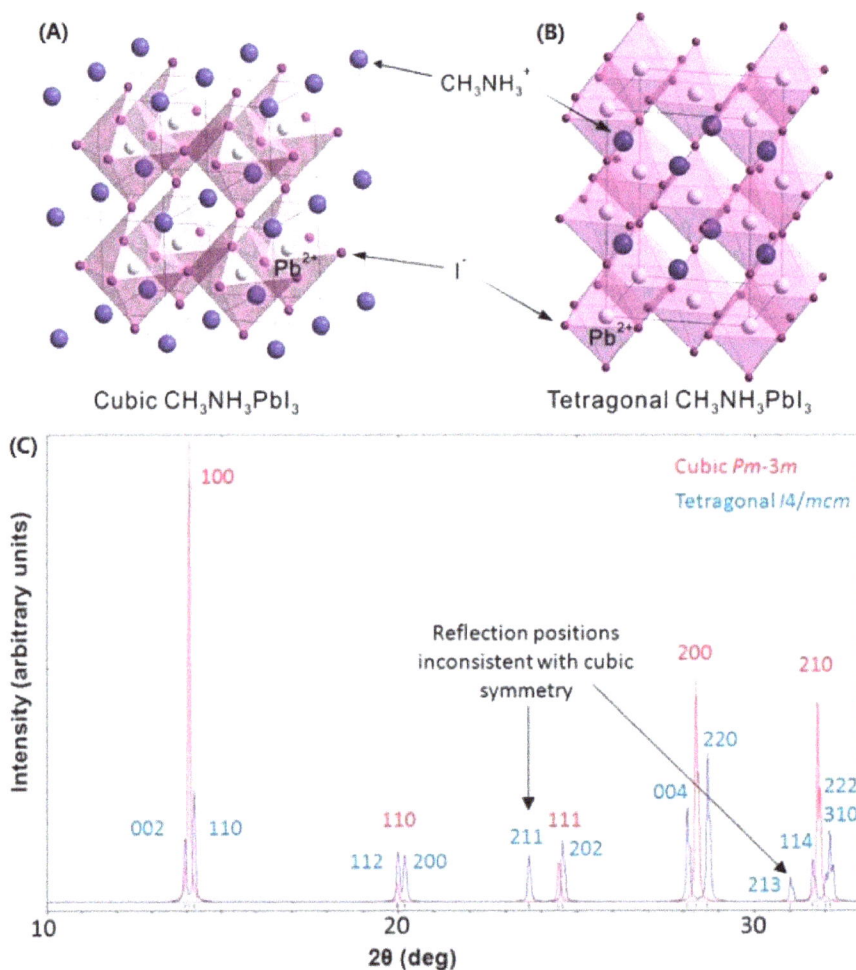

Figure 1. (**A**) Atomic models of MAPbI$_3$ with cubic phase; and (**B**) tetragonal phase; (**C**) the calculated XRD patterns for MAPbI$_3$ in both phases. (**A**) and (**B**) are reprinted from reference [64], Copyright © IOP Publishing. Reproduced with permission. All rights reserved; (**C**) is reprinted from reference [65], Copyright © 2013, Royal Society of Chemistry.

3.2. Converting Lead Halides to Perovskite

In the sequential deposition method, PbI_2 or/and $PbCl_2$ were first dissolved in a solvent. As PbI_2 crystal has a layered structure, DMF can intercalate into the PbI_2 interlayer space and screen PbI_2 via Pb-O bonding [71–73]. When DMF is intercalated, the XRD peak of the PbI_2 (001) plane red shifts from $14.8°$ to $7.94°$ [72,73]. The red-shift of this XRD peak to $9.17°$ also indicates the intercalation of DMSO [43]. While $PbCl_2$ doesn't possess a similar layered structure as PbI_2, its solubility is poor where $PbCl_2$ nanoparticles may only suspend in the solvent [28]. However, depositing a mixture of PbI_2 and $PbCl_2$ on the substrates result in a new PbICl phase [74], whose crystal structure is similar to $PbCl_2$ [75].

At the beginning of the reaction of PbI_2 and MAI, a predominant peak at (220) appeared (as shown in Figure 2B). In other words, the $MAPbI_3$ preferentially grows along (220) plane at first. The annealing process increases the long range crystalline order and results in the predominant (110) peak instead. Noting the (220) is only a short range of (110), thus, another possible reason for the (110)-oriented growth of $MAPbI_{3-x}Cl_x$ and $MAPbI_3$ may be because layered crystal structure of PbI_2 (growth along (001) planes of PbI_2 like the liquid catalyst cluster model mentioned in reference [76]). The lattice planes of tetragonal $MAPbI_3$ are showed in Figure 3.

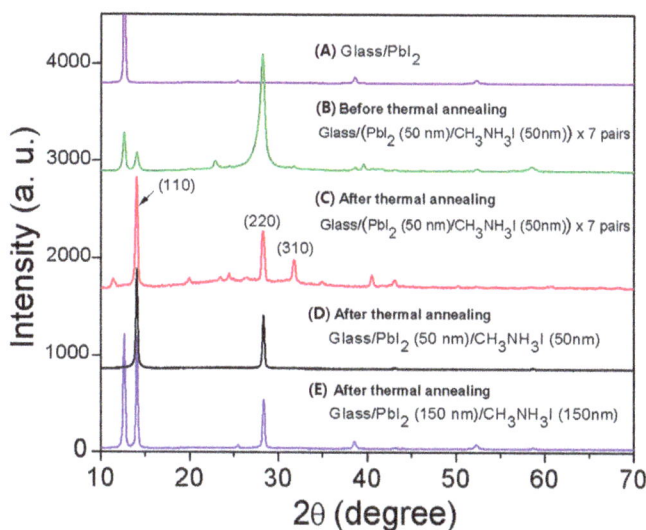

Figure 2. XRD patterns of glass substrates with vapor deposition of (**A**) a layer of PbI_2; and (**B**) a layer of PbI_2 followed by a layer of MAI, repeating this step 7 times; XRD pattern of the (**C**) annealed 7-time deposited PbI_2/MAI layer; (**D**) 1-time deposited PbI_2 (50 nm thickness)/MAI (50 nm thickness) layer; and (**E**) 1-time deposited PbI_2 (150 nm thickness)/MAI (150 nm thickness) layer. Reprinted from reference [77], Copyright © 2015, Royal Society of Chemistry.

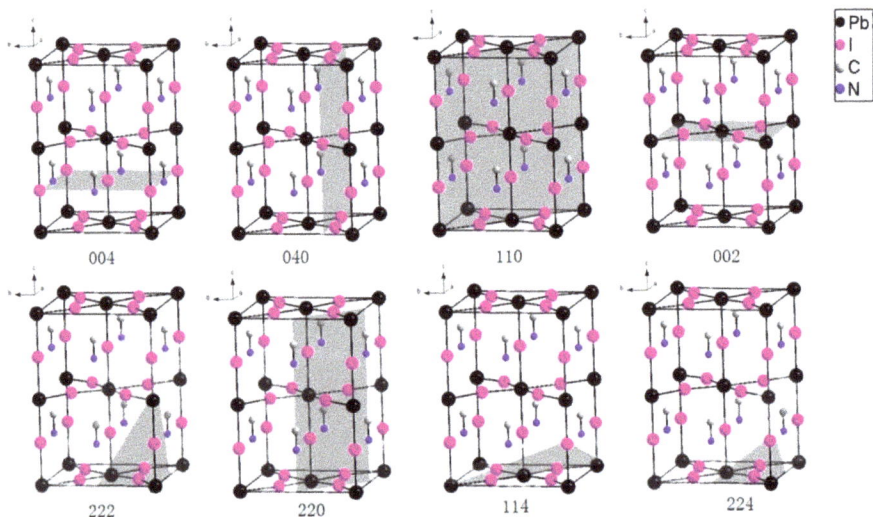

Figure 3. Crystallographic (lattice) planes (in gray) of tetragonal MAPbI3. Reprinted from reference [78], Copyright © 2015, American Chemical Society.

For $PbCl_2$, Cl^- was detached from $PbCl_2$ when the $PbCl_2$ was evaporated on the MAI substrate [79] and all the atoms of lead halide were dissociated during the crystal formation of the perovskite [80]. Thus, except the speed and the way of breaking the lead halide, the following steps should be similar with the one step method (Section 3.3) for converting PbI_2 or $PbCl_2$ with MAI to the perovskite. However, the situation in the presence of MACl may be different. As less energy is needed for MACl than MAI to undergo phase transition from solid to gas [69], it may be easier for MACl than MAI to diffuse into the PbI_2 and cause the crystallization of perovskite [81]. However, as Cl^- cannot be incorporated into $MAPbI_3$ crystal structure, the MAI and MACl may compete with each other to determine the result crystal, because only $MAPbI_3$ or $MAPbCl_3$ was formed when PbI2 was soaked in 80 mM MAI + 40 mM MACl or in 40 mM MAI + 80 mM MACl, respectively [80]. Thus, the incorporation of some amount of MACl managed to modify the morphology of the perovskite and resulted in better performance of the solar cells.

3.3. One Step Deposition of $MAPbI_{3-x}Cl_x$

The better crystallization of $MAPbI_{3-x}Cl_x$ along (110) and (220) plane of the tetragonal phase or (100) and (200) planes of the cubic phase may be due to the lowered cubic-tetragonal phase transition temperature of $MAPbI_{3-x}Cl_x$ after the incorporation of Cl^- [82]. A clear cubic-tetragonal phase transition temperature of $MAPbI_3$ was detected by differential scanning calorimeter (DSC) analysis [65], however no such phase transition was observed for $MAPbI_{3-x}Cl_x$ [83]. To explain the absence of the phase transition for MAPbI$_{3-x}$Cl$_x$, we first study the crystallization process of MAPbI$_{3-x}$Cl$_x$ by one step deposition method.

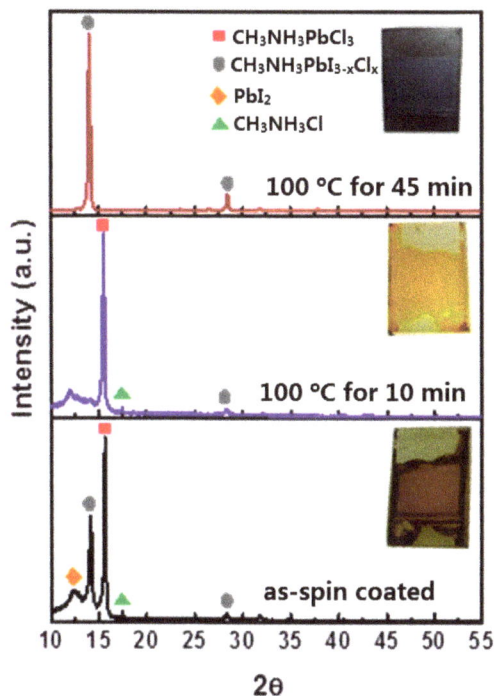

Figure 4. XRD patterns and optical images (insets) of MAPbI$_{3-x}$Cl$_x$ film during annealing. Reprinted from reference [84], Copyright © 2015, American Chemical Society.

Detail information about crystal formation process of MAPbI$_3$ is summarized in reference [85]. For MAPbI$_{3-x}$Cl$_x$, the transformation from the intermediate phase to the perovskite is determined as 80 °C by *in situ* grazing incidence wide-angle X-ray scattering (GIWAXS) [86]. Figure 4 presents a clearer picture of the crystal formation of MAPbI$_{3-x}$Cl$_x$. The 15.7° and 31.5° peaks are associated with the (100) and (200) diffraction peaks of MAPbCl$_3$ [82]. These peaks were also observed in references [27,87,88]. In Figure 4, it is interesting to note that MAPbI3 was formed first for the as-spin coated film but converted to MAPbCl$_3$ after annealing at 100 °C for 10 min, and then MAPbCl$_3$ was converted back to MAPbI$_3$ after 45 min of annealing [84]. Further annealing would result in the decomposition of MAPbI$_3$ to PbI$_2$, but this occurred after conversion to the intermediate phase to MAPbI$_3$ [89]. Because MAPbCl$_3$ is in a cubic phase, we suppose that MAPbCl$_3$ may cause a template effect for the cubic MAPbI$_3$ phase.

In addition, the MAI:PbI$_2$ (molar ratio 3:1) precursor solution on compact TiO$_2$ can also form MAPbI$_3$ with a predominant (110) plane, but the annealing temperature need to be above 150 °C [84,88,90]. The different sublimation temperature of MAI and MACl and the evidence of residue MAI or MACl in the resulting perovskite may explain the higher annealing temperature needed for MAPbI$_3$ [84,91].

The XRD patterns of the resulting MAPbI$_{3-x}$Cl$_x$ prepared from different chloride-containing precursors are summarized in Figure 5. All the patterns showed predominant crystallization along the (110) and (220) planes. Interestingly, the (220) peak split at a high MACl (x = 2) concentration in Figure 5B [27]. This split was also observed in reference [29]. In the sequential deposition method, the (110) and (220) crystallization preference may be due to an *in situ* transformation process [92] of PbI$_2$ to MAPbI$_3$, as discussed in Section 3.2. However, the PbI$_6$ octahedra are more likely to be fully dissociated in the one step precursor solution. [29,93–95] As MACl does not fit in the MAPbI$_3$ structure, it could be possible that MACl may be expelled along the (110) planes of the MAPbI$_3$ and that is why the MAPbI$_{3-x}$Cl$_x$ always showed the (110) and (220) orientation preference. This assumption can be proved by (220) peak split in Figure 5B, as excess of MACl breaks down the crystal range along (110) planes resulting in peak split. However, MAI can fit in the MAPbI$_3$ structure, (110)-oriented growth is just the result of cubic phase in high temperature (150 °C in refereneces [84,88,90]). Surprisingly, a main XRD peak of (310) was observed for the one step deposition prepared MAPbI$_3$ [96]. The main peak of (310), which is distinct from the (110) peak, may have resulted from the fact that the MAI was added into the precursor solution after the PbI$_2$ was completely dissolved instead of both MAI and PbI$_2$ being present at the same time [96], or the fact that the (310) plane of MAPbI$_3$ may match the crystal structure of the substrate better. Then the magnitude of the (110) peak of MAPbI$_{3-x}$Cl$_x$ and the (310) peak of MAPbI$_3$ further increases after 5 weeks [96]. Thus, we believe that the annealing process may only reinforce the crystallization preference as it is initially formed and the effects of substrate also contribute to the crystal structure formation of the perovskite in some cases. Returning to Figure 5, if excess of MACl breaks down the growth along the (110) plane, we believe MACl can also break down the crystalline order range. Since a large amount of MAPbI$_3$ existed in the amorphous phase form, the cubic phase of MAPbI$_3$ may be more favorable in short crystalline order range than the tetragonal phase.

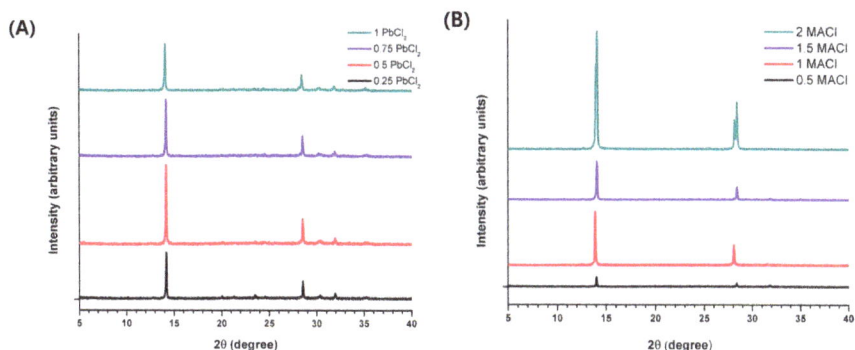

Figure 5. XRD patterns of MAPbI$_{3-x}$Cl$_x$ prepared from (**A**) precursor solution xPbCl$_2$+yPbI$_2$+zMAI (x = 0.25, 0.5, 0.75 and 1; y = 1 − x; z = 3 × x + y) in DMF; and (**B**) precursor solution 1PbI$_2$+1MAI+xMACl (x = 0.5, 1, 1.5 and 2). Reprinted from reference [27], Copyright © 2014, American Chemical Society.

There are other influences associated with Cl$^-$. Increasing the temperature during the soaking of the PbI$_2$ substrate in MAI + MACl IPA solution can improve the (110) orientation of MAPbI$_{3-x}$Cl$_x$ where the high temperature facilitates the expelling of MACl [97]. Annealing the MACl:PbI$_2$ (3:1) precursor on compact TiO$_2$ at 60 °C for 10 min followed by 100 °C for 20 min instead of gradually heating from 25 to 100 °C for 45 min resulted in the (200) crystal plane of MAPbI$_{3-x}$Cl$_x$ being vertically aligned on the substrate [98]. The tetragonal phase MAPbI$_{3-x}$Cl$_x$ was occasionally found on compact TiO$_2$ substrate [53], while the cubic phase always occurred in meso-porous substrate, where the trapped MACl in meso-porous structure [8] helps the formation of cubic phase. While the size of MAPbI$_3$ crystal grains are smaller but the degree of crystallinity improves in the presence of MACl [27,54], the sequential deposited MAPbI$_{3-x}$Cl$_x$ results in (001) elongated crystals [13].

4. Conclusions

In this article, the location of Cl$^-$ and its influence on the crystal morphology of MAPbI$_{3-x}$Cl$_x$ is summarized, where the deposition methods (one step deposition, sequential deposition and vapor based deposition) for MAPbI$_{3-x}$Cl$_x$ are reviewed. Furthermore, the cubic and tetragonal phases of MAPbI$_3$ are elucidated and the crystallization process of MAPbI$_{3-x}$Cl$_x$ is also summarized. Detailed information about the crystal structure with variable deposition parameters is also discussed. Though a recent report showed that Cl$^-$ mainly improves the carrier transport at the perovskite/Spiro-MeOTAD and perovskite/TiO$_2$ interfaces, rather than within the perovskite crystals, the authors of reference [99] more recently spatially resolved photoluminescence decay results showed less recombination in the high chlorine concentration region [100]. Thus, the effect of high concentration of Cl^- on the morphologies and electronic properties of the perovskite can still not be ignored. Additionally, whether Cl^- is predominantly present as a substituent for I^-, as an interstitial, or at the surface of the crystal, remains unclear [101] and this is worth further investigation.

Conflicts of Interest: Conflicts of Interest: The authors declare no conflict of interest.

References

1. Kojima, A.; Teshima, K.; Shirai, Y.; Miyasaka, T. Organometal halide perovskites as visible-light sensitizers for photovoltaic cells. *J. Am. Chem Soc.* **2009**, *131*, 6050–6051.
2. Yang, W.S.; Noh, J.H.; Jeon, N.J.; Kim, Y.C.; Ryu, S.; Seo, J.; Seok, S.I. High-performance photovoltaic perovskite layers fabricated through intramolecular exchange. *Science* **2015**, *348*, 1234–1237.
3. Kim, H.S.; Im, S.H.; Park, N.G. Organolead halide perovskite: New horizons in solar cell research. *J. Phys. Chem. C* **2014**, *118*, 5615–5625.
4. Luo, S.; Daoud, W.A. Recent progress in organic-inorganic halide perovskite solar cells: Mechanisms and materials design. *J. Mater. Chem. A* **2015**, *3*, 8992–9010.
5. Mohammad, K.N.; Gao, P.; Gratzel, M. Organohalide lead perovskites for photovoltaic applications. *Energ. Environ. Sci.* **2014**, *7*, 2448–2463.
6. Dualeh, A.; Tétreault, N.; Moehl, T.; Gao, P.; Nazeeruddin, M.K.; Grätzel, M. Effect of annealing temperature on film morphology of organic-inorganic hybrid pervoskite solid-state solar cells. *Adv. Funct. Mater.* **2014**, *24*, 3250–3258.
7. Stranks, S.D.; Eperon, G.E.; Grancini, G.; Menelaou, C.; Alcocer, M.J.P.; Leijtens, T.; Herz, L.M.; Petrozza, A.; Snaith, H.J. Electron-hole diffusion lengths exceeding 1 micrometer in an organometal trihalide perovskite absorber. *Science* **2013**, *342*, 341–344.
8. Colella, S.; Mosconi, E.; Fedeli, P.; Listorti, A.; Gazza, F.; Orlandi, F.; Ferro, P.; Besagni, T.; Rizzo, A.; Calestani, G.; *et al.* $MAPbI_{3-x}Cl_x$ mixed halide perovskite for hybrid solar cells: The role of chloride as dopant on the transport and structural properties. *Chem Mater.* **2013**, *25*, 4613–4618.
9. Conings, B.; Baeten, L.; de Dobbelaere, C.; D'Haen, J.; Manca, J.; Boyen, H.-G. Perovskite-based hybrid solar cells exceeding 10% efficiency with high reproducibility using a thin film sandwich approach. *Adv. Mater.* **2014**, *26*, 2041–2046.
10. Zhao, Y.; Zhu, K. CH_3NH_3Cl-assisted one-step solution growth of $CH_3NH_3PbI_3$: Structure, charge-carrier dynamics, and photovoltaic properties of perovskite solar cells. *J. Phys. Chem. C* **2014**, *118*, 9412–9418.
11. Colella, S.; Mosconi, E.; Pellegrino, G.; Alberti, A.; Guerra, V.L.P.; Masi, S.; Listorti, A.; Rizzo, A.; Condorelli, G.G.; De Angelis, F.; *et al.* Elusive presence of chloride in mixed halide perovskite solar cells. *J. Phys. Chem. Lett.* **2014**, *5*, 3532–3538.

12. Unger, E.L.; Bowring, A.R.; Tassone, C.J.; Pool, V.L.; Gold-Parker, A.; Cheacharoen, R.; Stone, K.H.; Hoke, E.T.; Toney, M.F.; McGehee, M.D. Chloride in lead chloride-derived organo-metal halides for perovskite-absorber solar cells. *Chem. Mater.* **2014**, *26*, 7158–7165.

13. Dar, M.I.; Arora, N.; Gao, P.; Ahmad, S.; Grätzel, M.; Nazeeruddin, M.K. Investigation regarding the role of chloride in organic-inorganic halide perovskites obtained from chloride containing precursors. *Nano Lett.* **2014**, *14*, 6991–6996.

14. Tripathi, N.; Yanagida, M.; Shirai, Y.; Masuda, T.; Han, L.; Miyano, K. Hysteresis-free and highly stable perovskite solar cells produced via a chlorine-mediated interdiffusion method. *J. Mater. Chem. A* **2015**, *3*, 12081–12088.

15. Starr, D.E.; Sadoughi, G.; Handick, E.; Wilks, R.G.; Alsmeier, J.H.; Kohler, L.; Gorgoi, M.; Snaith, H.J.; Bar, M. Direct observation of an inhomogeneous chlorine distribution in $CH_3NH_3PbI_{3-x}Cl_x$ layers: Surface depletion and interface enrichment. *Energ. Environ. Sci.* **2015**, *8*, 1609–1615.

16. Pool, V.L.; Gold-Parker, A.; McGehee, M.D.; Toney, M.F. Chlorine in $PbCl_2$-derived hybrid-perovskite solar absorbers. *Chem. Mater.* **2015**, *27*, 7240–7243.

17. Chae, J.; Dong, Q.; Huang, J.; Centrone, A. Chloride incorporation process in $CH_3NH_3PbI_{3-x}Cl_x$ perovskites via nanoscale bandgap maps. *Nano Lett.* **2015**, *15*, 8114–8121.

18. Mosconi, E.; Amat, A.; Nazeeruddin, M.K.; Grätzel, M.; De Angelis, F. First-principles modeling of mixed halide organometal perovskites for photovoltaic applications. *J. Phys. Chem. C* **2013**, *117*, 13902–13913.

19. Zheng, F.; Takenaka, H.; Wang, F.; Koocher, N.Z.; Rappe, A.M. First-principles calculation of the bulk photovoltaic effect in $CH_3NH_3PbI_3$ and $CH_3NH_3PbI_{3-x}Cl_x$. *J. Phys. Chem. Lett.* **2014**, *6*, 31–37.

20. Buin, A.; Pietsch, P.; Xu, J.; Voznyy, O.; Ip, A.H.; Comin, R.; Sargent, E.H. Materials processing routes to trap-free halide perovskites. *Nano Lett.* **2014**, *14*, 6281–6286.

21. Du, M.H. Efficient carrier transport in halide perovskites: Theoretical perspectives. *J. Mater. Chem. A* **2014**, *2*, 9091–9098.

22. Even, J.; Pedesseau, L.; Katan, C. Analysis of multivalley and multibandgap absorption and enhancement of free carriers related to exciton screening in hybrid perovskites. *J. Phys. Chem. C* **2014**, *118*, 11566–11572.

23. Yin, W.-J.; Shi, T.; Yan, Y. Unique properties of halide perovskites as possible origins of the superior solar cell performance. *Adv. Mater.* **2014**, *26*, 4653–4658.

24. Li, D.; Liang, C.; Zhang, H.; Zhang, C.; You, F.; He, Z. Spatially separated charge densities of electrons and holes in organic-inorganic halide perovskites. *J. Appl. Phys.* **2015**, *117*, 074901.

25. Yin, J.; Cortecchia, D.; Krishna, A.; Chen, S.; Mathews, N.; Grimsdale, A.C.; Soci, C. Interfacial charge transfer anisotropy in polycrystalline lead iodide perovskite films. *J. Phys. Chem. Lett.* **2015**, 1396–1402.

26. Mosconi, E.; Ronca, E.; De Angelis, F. First-principles investigation of the TiO_2/organohalide perovskites interface: The role of interfacial chlorine. *J. Phys. Chem. Lett.* **2014**, *5*, 2619–2625.

27. Williams, S.T.; Zuo, F.; Chueh, C.-C.; Liao, C.-Y.; Liang, P.-W.; Jen, A.K.Y. Role of chloride in the morphological evolution of organo-lead halide perovskite thin films. *Acs Nano* **2014**, *8*, 10640–10654.

28. Tidhar, Y.; Edri, E.; Weissman, H.; Zohar, D.; Hodes, G.; Cahen, D.; Rybtchinski, B.; Kirmayer, S. Crystallization of methyl ammonium lead halide perovskites: Implications for photovoltaic applications. *J. Am. Chem. Soc.* **2014**, *136*, 13249–13256.

29. Yan, K.; Long, M.; Zhang, T.; Wei, Z.; Chen, H.; Yang, S.; Xu, J. Hybrid halide perovskite solar cell precursors: Colloidal chemistry and coordination engineering behind device processing for high efficiency. *J. Am. Chem. Soc.* **2015**, *137*, 4460–4468.

30. Zheng, L.; Zhang, D.; Ma, Y.; Lu, Z.; Chen, Z.; Wang, S.; Xiao, L.; Gong, Q. Morphology control of the perovskite films for efficient solar cells. *Dalton Trans.* **2015**, *44*, 10582–10593.

31. Kim, H.S.; Lee, C.R.; Im, J.H.; Lee, K.B.; Moehl, T.; Marchioro, A.; Moon, S.J.; Humphry-Baker, R.; Yum, J.H.; Moser, J.E.; *et al.* Lead iodide perovskite sensitized all-solid-state submicron thin film mesoscopic solar cell with efficiency exceeding 9%. *Sci. Rep.* **2012**, *2*.

32. Lee, M.M.; Teuscher, J.; Miyasaka, T.; Murakami, T.N.; Snaith, H.J. Efficient hybrid solar cells based on meso-superstructured organometal halide perovskites. *Science* **2012**, *338*, 643–647.

33. Saliba, M.; Tan, K.W.; Sai, H.; Moore, D.T.; Scott, T.; Zhang, W.; Estroff, L.A.; Wiesner, U.; Snaith, H.J. Influence of thermal processing protocol upon the crystallization and photovoltaic performance of organic-inorganic lead trihalide perovskites. *J. Phys. Chem. C* **2014**, *118*, 17171–17177.

34. Hsu, H.-L.; Chen, C.; Chang, J.-Y.; Yu, Y.-Y.; Shen, Y.-K. Two-step thermal annealing improves the morphology of spin-coated films for highly efficient perovskite hybrid photovoltaics. *Nanoscale* **2014**, *6*, 10281–10288.

35. Kang, R.; Kim, J.-E.; Yeo, J.-S.; Lee, S.; Jeon, Y.-J.; Kim, D.-Y. Optimized organometal halide perovskite planar hybrid solar cells via control of solvent evaporation rate. *J. Phys. Chem. C* **2014**, *118*, 26513–26520.

36. Guo, Y.; Liu, C.; Inoue, K.; Harano, K.; Tanaka, H.; Nakamura, E. Enhancement in the efficiency of an organic-inorganic hybrid solar cell with a doped P3HT hole-transporting layer on a void-free perovskite active layer. *J. Mater. Chem. A* **2014**, *2*, 13827–13830.

37. Liang, P.-W.; Liao, C.-Y.; Chueh, C.-C.; Zuo, F.; Williams, S.T.; Xin, X.-K.; Lin, J.; Jen, A.K.Y. Additive enhanced crystallization of solution-processed perovskite for highly efficient planar-heterojunction solar cells. *Adv. Mater.* **2014**, *26*, 3748–3754.

38. Chueh, C.-C.; Liao, C.-Y.; Zuo, F.; Williams, S.T.; Liang, P.-W.; Jen, A.K.Y. The roles of alkyl halide additives in enhancing perovskite solar cell performance. *J. Mater. Chem. A* **2014**, *3*, 9058–9062.

39. Ding, Y.; Yao, X.; Zhang, X.; Wei, C.; Zhao, Y. Surfactant enhanced surface coverage of $CH_3NH_3PbI_{3-x}Cl_x$ perovskite for highly efficient mesoscopic solar cells. *J. Power Sources* **2014**, *272*, 351–355.

40. Eperon, G.E.; Burlakov, V.M.; Docampo, P.; Goriely, A.; Snaith, H.J. Morphological control for high performance, solution-processed planar heterojunction perovskite solar cells. *Adv. Funct. Mater.* **2014**, *24*, 151–157.

41. Takeo, O.; Masahito, Z.; Yuma, I.; Atsushi, S.; Kohei, S. Microstructures and photovoltaic properties of perovskite-type $CH_3NH_3PbI_3$ compounds. *Appl. Phys. Express.* **2014**, *7*.

42. Jeon, Y.-J.; Lee, S.; Kang, R.; Kim, J.-E.; Yeo, J.-S.; Lee, S.-H.; Kim, S.-S.; Yun, J.-M.; Kim, D.-Y. Planar heterojunction perovskite solar cells with superior reproducibility. *Sci. Rep.* **2014**, *4*.

43. Jeon, N.J.; Noh, J.H.; Kim, Y.C.; Yang, W.S.; Ryu, S.; Seok, S.I. Solvent engineering for high-performance inorganic-organic hybrid perovskite solar cells. *Nat. Mater.* **2014**, *13*, 897–903.

44. Xiao, M.; Huang, F.; Huang, W.; Dkhissi, Y.; Zhu, Y.; Etheridge, J.; Gray-Weale, A.; Bach, U.; Cheng, Y.-B.; Spiccia, L. A fast deposition-crystallization procedure for highly efficient lead iodide perovskite thin-film solar cells. *Angew. Chem.* **2014**, *126*, 10056–10061.

45. Jung, J.W.; Williams, S.T.; Jen, A.K.Y. Low-temperature processed high-performance flexible perovskite solar cells via rationally optimized solvent washing treatments. *Rsc Adv.* **2014**, *4*, 62971–62977.

46. Ito, S.; Tanaka, S.; Vahlman, H.; Nishino, H.; Manabe, K.; Lund, P. Carbon-double-bond-free printed solar cells from $TiO_2/CH_3NH_3PbI_3/CuSCN/Au$: Structural control and photoaging effects. *Chem. Phys. Chem.* **2014**, *15*, 1194–1200.

47. Huang, F.; Dkhissi, Y.; Huang, W.; Xiao, M.; Benesperi, I.; Rubanov, S.; Zhu, Y.; Lin, X.; Jiang, L.; Zhou, Y.; *et al.* Gas-assisted preparation of lead iodide perovskite films consisting of a monolayer of single crystalline grains for high efficiency planar solar cells. *Nano Energy* **2014**, *10*, 10–18.

48. Burschka, J.; Pellet, N.; Moon, S.J.; Humphry-Baker, R.; Gao, P.; Nazeeruddin, M.K.; Gratzel, M. Sequential deposition as a route to high-performance perovskite-sensitized solar cells. *Nature* **2013**, *499*, 316–320.

49. Ma, Y.; Zheng, L.; Chung, Y.-H.; Chu, S.; Xiao, L.; Chen, Z.; Wang, S.; Qu, B.; Gong, Q.; Wu, Z.; *et al.* A highly efficient mesoscopic solar cell based on $CH_3NH_3PbI_{3-x}Cl_x$ fabricated via sequential solution deposition. *Chem. Commun.* **2014**, *50*, 12458–12461.

50. Zheng, L.; Ma, Y.; Chu, S.; Wang, S.; Qu, B.; Xiao, L.; Chen, Z.; Gong, Q.; Wu, Z.; Hou, X. Improved light absorption and charge transport for perovskite solar cells with rough interfaces by sequential deposition. *Nanoscale* **2014**, *6*, 8171–8176.

51. Dar, M.I.; Abdi-Jalebi, M.; Arora, N.; Grätzel, M.; Nazeeruddin, M.K. Growth engineering of $CH_3NH_3PbI_3$ structures for high-efficiency solar cells. *Adv. Energy Mater.* **2016**, *6*.

52. Li, Y.; Sun, W.; Yan, W.; Ye, S.; Peng, H.; Liu, Z.; Bian, Z.; Huang, C. High-performance planar solar cells based on $CH_3NH_3PbI_{3-x}Cl_x$ perovskites with determined chlorine mole fraction. *Adv. Funct. Mater.* **2015**, *25*, 4867–4873.

53. Docampo, P.; Hanusch, F.; Stranks, S.D.; Döblinger, M.; Feckl, J.M.; Ehrensperger, M.; Minar, N.K.; Johnston, M.B.; Snaith, H.J.; Bein, T. Solution deposition-conversion for planar heterojunction mixed halide perovskite solar cells. *Adv. Energy Mater.* **2014**, *4*.

54. Jiang, M.; Wu, J.; Lan, F.; Tao, Q.; Gao, D.; Li, G. Enhancing the performance of planar organo-lead halide perovskite solar cells by using mixed halide source. *J. Mater. Chem. A* **2014**, *3*, 963–967.

55. Dong, Q.; Yuan, Y.; Shao, Y.; Fang, Y.; Wang, Q.; Huang, J. Abnormal crystal growth in $CH_3NH_3PbI_{3-x}Cl_x$ using a multi-cycle solution coating process. *Energ. Environ. Sci.* **2015**, *8*, 2464–2470.

56. Liu, M.; Johnston, M.B.; Snaith, H.J. Efficient planar heterojunction perovskite solar cells by vapour deposition. *Nature* **2013**, *501*, 395–398.

57. Ono, L.K.; Wang, S.; Kato, Y.; Raga, S.R.; Qi, Y. Fabrication of semi-transparent perovskite films with centimeter-scale superior uniformity by the hybrid deposition method. *Energ. Environ. Sci.* **2014**, *7*, 3989–3993.

58. Leyden, M.R.; Ono, L.K.; Raga, S.R.; Kato, Y.; Wang, S.; Qi, Y. High performance perovskite solar cells by hybrid chemical vapor deposition. *J. Mater. Chem. A* **2014**, *2*, 18742–18745.

59. Chen, C.-W.; Kang, H.-W.; Hsiao, S.-Y.; Yang, P.-F.; Chiang, K.-M.; Lin, H.-W. Efficient and uniform planar-type perovskite solar cells by simple sequential vacuum deposition. *Adv. Mater.* **2014**, *26*, 6647–6652.

60. Poglitsch, A.; Weber, D. Dynamic disorder in methylammoniumtrihalogenoplumbates (ii) observed by millimeter-wave spectroscopy. *J. Chem. Phys.* **1987**, *87*, 6373–6378.

61. Stoumpos, C.C.; Malliakas, C.D.; Kanatzidis, M.G. Semiconducting tin and lead iodide perovskites with organic cations: Phase transitions, high mobilities, and near-infrared photoluminescent properties. *Inorg. Chem.* **2013**, *52*, 9019–9038.

62. Kawamura, Y.; Mashiyama, H.; Hasebe, K. Structural study on cubic–tetragonal transition of $CH_3NH_3PbI_3$. *J. Phy. Soc. Jpn.* **2002**, *71*, 1694–1697.

63. Eperon, G.E.; Stranks, S.D.; Menelaou, C.; Johnston, M.B.; Herz, L.M.; Snaith, H.J. Formamidinium lead trihalide: A broadly tunable perovskite for efficient planar heterojunction solar cells. *Energ. Environ. Sci.* **2014**, *7*, 982–988.

64. Chen, Z.; Li, H.; Tang, Y.; Huang, X.; Ho, D.; Lee, C.-S. Shape-controlled synthesis of organolead halide perovskite nanocrystals and their tunable optical absorption. *Mater. Res. Express* **2014**, *1*.

65. Baikie, T.; Fang, Y.N.; Kadro, J.M.; Schreyer, M.; Wei, F.X.; Mhaisalkar, S.G.; Graetzel, M.; White, T.J. Synthesis and crystal chemistry of the hybrid perovskite $CH_3NH_3PbI_3$ for solid-state sensitised solar cell applications. *J. Mater. Chem. A* **2013**, *1*, 5628–5641.

66. Choi, J.J.; Yang, X.H.; Norman, Z.M.; Billinge, S.J.L.; Owen, J.S. Structure of methylammonium lead iodide within mesoporous titanium dioxide: Active material in high-performance perovskite solar cells. *Nano Lett.* **2014**, *14*, 127–133.

67. Zhou, Y.; Vasiliev, A.L.; Wu, W.; Yang, M.; Pang, S.; Zhu, K.; Padture, N.P. Crystal morphologies of organolead trihalide in mesoscopic/planar perovskite solar cells. *J. Phys. Chem. Lett.* **2015**, *6*, 2292–2297.

68. Harms, H.A.; Tetreault, N.; Pellet, N.; Bensimon, M.; Gratzel, M. Mesoscopic photosystems for solar light harvesting and conversion: Facile and reversible transformation of metal-halide perovskites. *Faraday Discuss.* **2014**, *176*, 251–269.

69. Dualeh, A.; Gao, P.; Seok, S.I.; Nazeeruddin, M.K.; Grätzel, M. Thermal behavior of methylammonium lead-trihalide perovskite photovoltaic light harvesters. *Chem Mater.* **2014**, *26*, 6160–6164.

70. Park, B.-W.; Jain, S.M.; Zhang, X.; Hagfeldt, A.; Boschloo, G.; Edvinsson, T. Resonance raman and excitation energy dependent charge transfer mechanism in halide-substituted hybrid perovskite solar cells. *Acs Nano* **2015**, *9*, 2088–2101.

71. Wakamiya, A.; Endo, M.; Sasamori, T.; Tokitoh, N.; Ogomi, Y.; Hayase, S.; Murata, Y. Reproducible fabrication of efficient perovskite-based solar cells: X-ray crystallographic studies on the formation of $CH_3NH_3PbI_3$ layers. *Chem. Lett.* **2014**, *43*, 711–713.

72. Hao, F.; Stoumpos, C.C.; Liu, Z.; Chang, R.P.H.; Kanatzidis, M.G. Controllable perovskite crystallization at a gas-solid interface for hole conductor-free solar cells with steady power conversion efficiency over 10%. *J. Am. Chem. Soc.* **2014**, *136*, 16411–16419.

73. Shen, D.; Yu, X.; Cai, X.; Peng, M.; Ma, Y.; Su, X.; Xiao, L.; Zou, D. Understanding the solvent-assisted crystallization mechanism inherent in efficient organic-inorganic halide perovskite solar cells. *J. Mater. Chem. A* **2014**, *2*, 20454–20461.

74. Li, Y.; Cooper, J.K.; Buonsanti, R.; Giannini, C.; Liu, Y.; Toma, F.M.; Sharp, I.D. Fabrication of planar heterojunction perovskite solar cells by controlled low-pressure vapor annealing. *J. Phys. Chem. Lett.* **2015**, *6*, 493–499.

75. Brixner, L.H.; Chen, H.Y.; Foris, C.M. X-ray study of the $PbCl_{2-x}I_x$ and $PbBr_{2-x}I_x$ systems. *J. Solid State Chem.* **1981**, *40*, 336–343.

76. Im, J.-H.; Luo, J.; Franckevičius, M.; Pellet, N.; Gao, P.; Moehl, T.; Zakeeruddin, S.M.; Nazeeruddin, M.K.; Grätzel, M.; Park, N.-G. Nanowire perovskite solar cell. *Nano Lett.* **2015**, *15*, 2120–2126.

77. Ng, A.; Ren, Z.; Shen, Q.; Cheung, S.H.; Gokkaya, H.C.; Bai, G.; Wang, J.; Yang, L.; So, S.K.; Djurisic, A.B.; *et al.* Efficiency enhancement by defect engineering in perovskite photovoltaic cells prepared using evaporated PbI_2/CH_3NH_3I multilayers. *J. Mater. Chem. A* **2015**, *3*, 9223–9231.

78. Zhu, F.; Men, L.; Guo, Y.; Zhu, Q.; Bhattacharjee, U.; Goodwin, P.M.; Petrich, J.W.; Smith, E.A.; Vela, J. Shape evolution and single particle luminescence of organometal halide perovskite nanocrystals. *Acs Nano* **2015**, *9*, 2948–2959.

79. Ng, T.-W.; Chan, C.-Y.; Lo, M.-F.; Guan, Z.Q.; Lee, C.-S. Formation chemistry of perovskites with mixed iodide/chloride content and the implications on charge transport properties. *J. Mater. Chem. A* **2015**, *3*, 9081–9085.

80. Moore, D.T.; Sai, H.; Tan, W.K.; Estroff, L.A.; Wiesner, U. Impact of the organic halide salt on final perovskite composition for photovoltaic applications. *APL Mater.* **2014**, *2*.

81. Xu, Y.; Zhu, L.; Shi, J.; Lv, S.; Xu, X.; Xiao, J.; Dong, J.; Wu, H.; Luo, Y.; Li, D.; *et al.* Efficient hybrid mesoscopic solar cells with morphology-controlled $CH_3NH_3PbI_{3-x}Cl_x$ derived from two-step spin coating method. *Acs Appl. Mater. Inter.* **2015**, *7*, 2242–2248.

82. Pistor, P.; Borchert, J.; Fränzel, W.; Csuk, R.; Scheer, R. Monitoring the phase formation of co-evaporated lead halide perovskite thin films by *in situ* XRD. *J. Phys. Chem. Lett.* **2014**, *5*, 3308–3312.

83. Williams, A.E.; Holliman, P.J.; Carnie, M.J.; Davies, M.L.; Worsley, D.A.; Watson, T.M. Perovskite processing for photovoltaics: A spectro-thermal evaluation. *J. Mater. Chem. A* **2014**, *2*, 19338–19346.

84. Yantara, N.; Yanan, F.; Shi, C.; Dewi, H.A.; Boix, P.P.; Mhaisalkar, S.G.; Mathews, N. Unravelling the effects of Cl addition in single step $CH_3NH_3PbI_3$ perovskite solar cells. *Chem. Mater.* **2015**, *27*, 2309–2314.

85. Song, Z.; Watthage, S.C.; Phillips, A.B.; Tompkins, B.L.; Ellingson, R.J.; Heben, M.J. Impact of processing temperature and composition on the formation of methylammonium lead iodide perovskites. *Chem. Mater.* **2015**, *27*, 4612–4619.

86. Tan, K.W.; Moore, D.T.; Saliba, M.; Sai, H.; Estroff, L.A.; Hanrath, T.; Snaith, H.J.; Wiesner, U. Thermally induced structural evolution and performance of mesoporous block copolymer-directed alumina perovskite solar cells. *Acs Nano* **2014**, *8*, 4730–4739.

87. Zhou, H.; Chen, Q.; Li, G.; Luo, S.; Song, T.-b.; Duan, H.-S.; Hong, Z.; You, J.; Liu, Y.; Yang, Y. Interface engineering of highly efficient perovskite solar cells. *Science* **2014**, *345*, 542–546.

88. Yu, H.; Wang, F.; Xie, F.; Li, W.; Chen, J.; Zhao, N. The role of chlorine in the formation process of "$CH_3NH_3PbI_{3-x}Cl_x$" perovskite. *Adv. Funct. Mater.* **2014**, *24*, 7102–7108.

89. Song, T.-B.; Chen, Q.; Zhou, H.; Luo, S.; Yang, Y.; You, J. Unraveling film transformations and device performance of planar perovskite solar cells. *Nano Energy* **2015**, *12*, 494–500.

90. Grancini, G.; Marras, S.; Prato, M.; Giannini, C.; Quarti, C.; De Angelis, F.; De Bastiani, M.; Eperon, G.E.; Snaith, H.J.; Manna, L.; *et al.* The impact of the crystallization processes on the structural and optical properties of hybrid perovskite films for photovoltaics. *J. Phys. Chem. Lett.* **2014**, *5*, 3836–3842.

91. Pathak, S.; Sepe, A.; Sadhanala, A.; Deschler, F.; Haghighirad, A.; Sakai, N.; Goedel, K.C.; Stranks, S.D.; Noel, N.; Price, M.; *et al.* Atmospheric influence upon crystallization and electronic disorder and its impact on the photophysical properties of organic-inorganic perovskite solar cells. *Acs Nano* **2015**, *9*, 2311–2320.

92. Yang, S.; Zheng, Y.C.; Hou, Y.; Chen, X.; Chen, Y.; Wang, Y.; Zhao, H.; Yang, H.G. Formation mechanism of freestanding $CH_3NH_3PbI_3$ functional crystals: *In situ* transformation vs dissolution-crystallization. *Chem. Mater.* **2014**, *26*, 6705–6710.

93. Shkrob, I.A.; Marin, T.W. Charge trapping in photovoltaically active perovskites and related halogenoplumbate compounds. *J. Phys. Chem. Lett* **2014**, *5*, 1066–1071.

94. Wang, Q.; Yun, J.-H.; Zhang, M.; Chen, H.; Chen, Z.-G.; Wang, L. Insight into the liquid state of organo-lead halide perovskites and their new roles in dye-sensitized solar cells. *J. Mater. Chem A* **2014**, *2*, 10355–10358.

95. Stamplecoskie, K.G.; Manser, J.S.; Kamat, P.V. Dual nature of the excited state in organic-inorganic lead halide perovskites. *Energ. Environ. Sci.* **2015**, *8*, 208–215.

96. Park, B.-W.; Philippe, B.; Gustafsson, T.; Sveinbjörnsson, K.; Hagfeldt, A.; Johansson, E.M.J.; Boschloo, G. Enhanced crystallinity in organic-inorganic lead halide perovskites on mesoporous Tio$_2$ via disorder-order phase transition. *Chem. Mater.* **2014**, *26*, 4466–4471.

97. Docampo, P.; Hanusch, F.C.; Giesbrecht, N.; Angloher, P.; Ivanova, A.; Bein, T. Influence of the orientation of methylammonium lead iodide perovskite crystals on solar cell performance. *APL Mater.* **2014**, *2*.

98. Ishii, A.; Jena, A.K.; Miyasaka, T. Fully crystalline perovskite-perylene hybrid photovoltaic cell capable of 1.2 V output with a minimized voltage lossa. *APL Mater.* **2014**, *2*.

99. Chen, Q.; Zhou, H.; Fang, Y.; Stieg, A.Z.; Song, T.-B.; Wang, H.-H.; Xu, X.; Liu, Y.; Lu, S.; You, J.; *et al.* The optoelectronic role of chlorine in CH$_3$NH$_3$PbI$_3$(Cl)-based perovskite solar cells. *Nat. Commun.* **2015**, *6*.

100. De Quilettes, D.W.; Vorpahl, S.M.; Stranks, S.D.; Nagaoka, H.; Eperon, G.E.; Ziffer, M.E.; Snaith, H.J.; Ginger, D.S. Impact of microstructure on local carrier lifetime in perovskite solar cells. *Science* **2015**, *348*, 683–686.

101. Stranks, S.D.; Nayak, P.K.; Zhang, W.; Stergiopoulos, T.; Snaith, H.J. Formation of thin films of organic-inorganic perovskites for high-efficiency solar cells. *Angew. Chem. Int. Ed.* **2015**, *54*, 3240–3248.

Multi-Material Front Contact for 19% Thin Film Solar Cells

Joop van Deelen, Yasemin Tezsevin and Marco Barink

Abstract: The trade-off between transmittance and conductivity of the front contact material poses a bottleneck for thin film solar panels. Normally, the front contact material is a metal oxide and the optimal cell configuration and panel efficiency were determined for various band gap materials, representing $Cu(In,Ga)Se_2$ (CIGS), CdTe and high band gap perovskites. Supplementing the metal oxide with a metallic copper grid improves the performance of the front contact and aims to increase the efficiency. Various front contact designs with and without a metallic finger grid were calculated with a variation of the transparent conductive oxide (TCO) sheet resistance, scribing area, cell length, and finger dimensions. In addition, the contact resistance and illumination power were also assessed and the optimal thin film solar panel design was determined. Adding a metallic finger grid on a TCO gives a higher solar cell efficiency and this also enables longer cell lengths. However, contact resistance between the metal and the TCO material can reduce the efficiency benefit somewhat.

Reprinted from *Materials*. Cite as: van Deelen, J.; Tezsevin, Y.; Barink, M. Multi-Material Front Contact for 19% Thin Film Solar Cells. *Materials* **2016**, *9*, 96.

1. Introduction

Photovoltaics (PV) is a wide arena for materials science to demonstrate the power of bringing different materials together in one device. There are two main material based photovoltaic families: one is Si wafer based and the other is thin film PV, which relies on coating of high quality materials on a substrate [1]. Even though thin film PV is based on "simple" coating steps, the solar power conversion efficiency has been improved to values well above 20% and approaches the values previously only reached by Si record cells [2,3]. One of the main drivers behind this success is material improvement. In addition, interface issues have been tackled. For instance, recently back surface passivation in CIGS cells has been improved [4,5]. However, unlike for Si wafer based technologies, the stunning laboratory cell advances have not translated into 20% solar panel efficiencies.

Two of the bottlenecks for thin film solar panels are the active area loss due to interconnection and losses in the transparent front contact, for which usually a transparent conductive oxide (TCO) is coated [6,7]. The loss in active area should be reduced and we detail its impact on overall cell and front contact design in the Results Section. The TCO inevitably has a trade-off between conductivity and transparency [8,9]. In the case of small cells, this hurdle can be masked using small dimensions and addition of a patterned metallic grid to the TCO, thereby compensating for its low conductivity. Such a combination of materials can increase the efficiency by creating a dramatic shift in conductivity at the expense of only a small loss in transmittance [10,11].

Classic wafer based cells do not have a TCO and fully rely on metallic grids. Therefore, the cell layout and ink requirements are highly different from the desired characteristics of monolithically interconnected thin film cells, which requires smaller feature sizes and, in the case of thin film CIGS cells, limited annealing temperatures below 200 °C. A few studies of grid on TCO were performed, but these reflected the status of ink jet printing, resulting in low (<1 μm) and 100 μm wide grids. Because of these low and wide grid dimensions, there was hardly an advantage of this TCO + metal grid approach compared to the TCO only approach [12–15]. Therefore, in solar panel production, this solution has not been adopted for monolithically interconnected solar cells.

Recently, however, the full potential of the application of metallic grids with optimized finger and cell dimensions (*i.e.*, lower width and larger height) was reported to give a significant boost in thin film solar efficiencies [16]. Because such approach would add complexity in the manufacturing process, the efficiency gain should be determined and evaluated with respect to manufacturing issues. Previous study [16,17] was performed on cells with efficiencies of 15.5% and many aspects such as cell layout and absorber material band gap, were not discussed in depth. Moreover, previous designs did not take into account the material interface issue of contact resistance. The investigation of contact resistance in solar cells has only briefly been touched [18,19] and its impact on design of monolithically integrated solar panels still needs to be addressed. Moreover, the previous case was limited to low efficiency CIGS or organic PV [15,16] and the case for high efficiency thin film solar cells, spanning a wide range of band gaps should be investigated. In short, there is a lack of knowledge of the impact of the cell layout and specific metal-TCO interaction (e.g., contact resistance) on the preferred grid design and the expected efficiency benefit.

This work focuses on the introduction of metal finger grid to enhance the performance of thin film solar panels with up-to-date cell efficiencies of 19%. The effects of cell length and interconnection area, as well as the band gap of the absorber material and the contact resistance are modeled. In contrast to previous work reflecting a rather idealized situation, specific issues such as the losses due to the specific contact resistance and the impact of reduced irradiation intensity are discussed. The calculated cell efficiencies give guidelines over a wide range of (non-ideal) circumstances for useful front contact technologies that aim to enhance the thin film solar panel efficiency.

2. Results and Discussion

2.1. General Considerations

In thin film solar panels, the panel is usually divided into parallel cells that are series connected. There are several ways to accomplish this and Figure 1 details two of them.

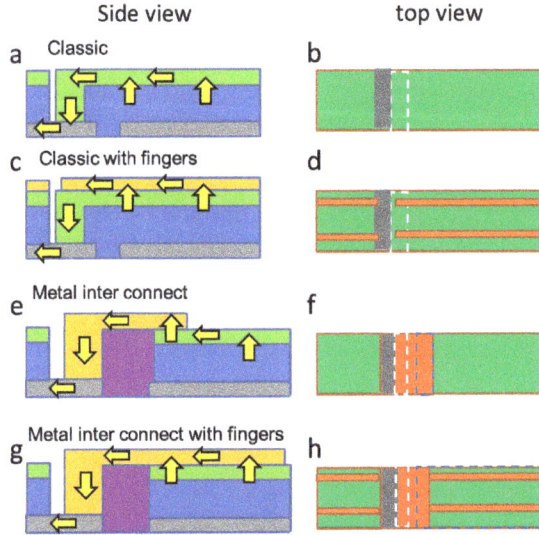

Figure 1. Schematic representations (not to scale) of different interconnection and cell layouts with a side view (**a,c,e,g**) and a top view (**b,d,f,h**). The top image shows the front contact (in green), the absorber material (in blue) and the back contact (in grey). In addition, the separation and interconnection layout between two adjacent cells is shown. The surface area of the TCO/back contact material interface is indicated by the white dashed box. The flow of current is depicted by the arrows. The second highest image shows the case where the front contact is supplemented by a metal grid (in orange), whereas the right image displays the area covered by the metal (not to scale). The third image shows the case of the metal interconnect, for which two material interfaces are important: the metal back contact area represented by the white dashed box and the metal/TCO contact areas represented by the blue dashed box.

The first is the classic way, in which the TCO is both the front contact and the interconnect (Figure 1a,b). In this case, the isolation area of the back contact is filled with the semiconducting absorber material and all the current is transported through the TCO. The TCO can be enhanced by a metallic finger grid, while the interconnection between top and bottom electrode takes place at the TCO back

116

contact interface (Figure 1c,d). Alternatively, a metal busbar can be used for this interconnection purpose (Figure 1e,f) [20]. The isolation of the back contact can be filled with a dedicated insulator material [21]. This approach was mentioned to have more design freedom. In addition, the metal can function both as an interconnection and as a top contact enhancer (Figure 1g,h). The contact surface area between the front and the back contact, as indicated by the white dashed box, is not changed by these different layouts. In the case of the metal interconnect, the fingers on the TCO will increase the total contact surface area between the metal and the TCO, which is an important feature, as will be discussed in Section 2.4.

A modest cell efficiency of 19% was chosen, as this has been reported for different thin film materials with various band gaps, which result in different open circuit voltages. Three I-V curves were chosen with an efficiency of 19% and open circuit voltages (Voc) of 0.7, 0.9 and 1.1 V, as to represent typical values for thin film CIGS, CdTe and perovskite solar cells, respectively. The curves are shown in Figure 2a. More details of the IV curves can be found in the Experimental Section.

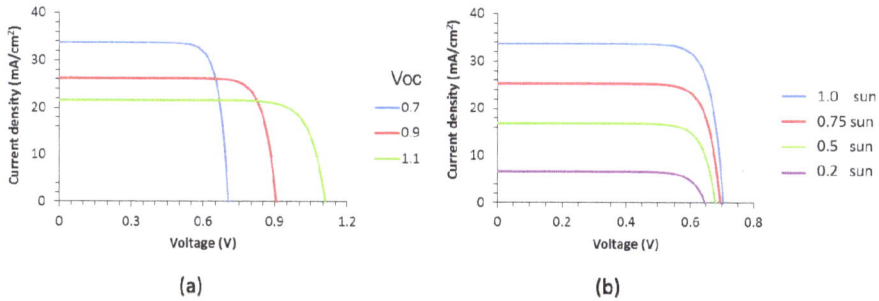

Figure 2. I-V characteristics used for the study (**a**) cells of 19% efficiency with different open circuit voltages (in V, see legend); and (**b**) cell with a Voc of 0.7 V for different light intensities (see legend) in which one sun is equivalent to $1000 \, W/m^2$).

For the curve with a Voc of 0.7 V, the illumination intensity was varied and its effect on the IV curve is shown in Figure 2b. As the light induced current density goes down, so do the Voc and the fill factor. These curves were used in the modeling to represent reference small cell without interconnection of front contact related losses.

2.2. Cells with a TCO Front Contact

The typical trade-off between transmittance and sheet resistance of the TCO, as used for the modeling, is shown in Figure 3. Below 10 Ω/sq, the transmittance drops with reduced sheet resistance. Figure 4a shows the efficiency as a function of the cell length for different TCO sheet resistances. The details of the TCO can be found in the Experimental Section. The cell efficiency shows a maximum with cell length.

For very short cells, the optical loss related to the scribing width that is needed for isolation and interconnection is high (here taken to be 150 μm, which is near the lowest reported for CIGS [22]) compared to the total cell area. For longer cells, the efficiency drops as resistive losses become a major bottleneck. Naturally, a TCO with a lower sheet resistance allows for longer cells. However, as a lower sheet resistance goes together with a lower TCO transmittance [23,24], as shown in Figure 3, there is a trade-off and as is obvious from Figure 4a, different TCO sheet resistance have a different optimal cell length. A TCO sheet resistance of 5 Ω/sq has a long optimal cell length, but as the transmittance TCO is substantially lower than that of 10 Ω/sq, the efficiency drops from 16.9 % to 16.2%.

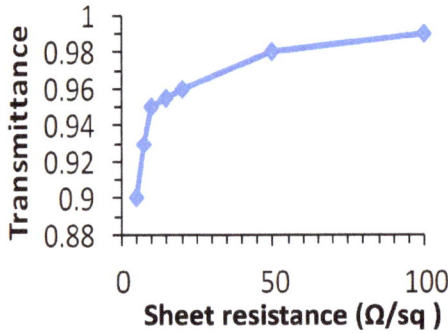

Figure 3. Transmittance as a function of the sheet resistance. This is used to represent TCO induced optical losses in industrially sputtered ZnO: Al material for a wavelength between 400 nm and 1100 nm and do not reflect state of the art laboratory results.

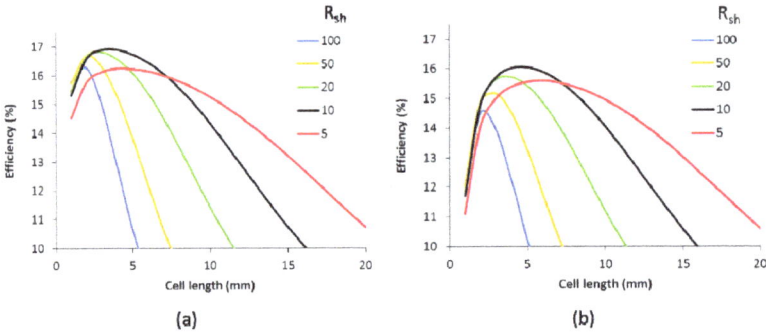

Figure 4. Efficiency of solar panels as a function of the individual cell length for different sheet resistances of the TCO (Rsh in Ω/sq) for a scribe width of 150 μm (**a**) and 350 μm (**b**). The cell was based on a Voc of 0.7 V.

A maximum efficiency of 16.9% is reached for a TCO sheet resistance of 10 Ω/sq. In other words, when going from a 19% small cell to a solar panel, scribing losses and TCO related losses reduce the panel efficiency by as much as 2 absolute %. When the scribing width is enlarged to 350 µm, which is now a common value in production, the maximum obtainable cell efficiency drops to 16%, as shown in Figure 4b. This indicates the importance of careful process control and the gain that can be obtained when material removal is more carefully controlled. Moreover, the maximum cell efficiency is obtained at slightly higher cell length, but this difference is rather small. Interestingly, the difference of maximum efficiencies between the high TCOs sheet resistances is increased. This can be explained as follow: a high sheet resistance requires short cells. As the wider scribing width translates to a larger sensitivity to more narrow cells, the impact will be higher.

Figure 5 demonstrates that high band gap materials with higher Voc translate in higher panel efficiencies, even though the small cell efficiency remains 19%. This can be explained by the fact that a higher Voc comes together with a lower short circuit density. This combination brings lower resistive losses. Moreover, lower resistive losses enable longer cells, which help to reduce the optical losses by the scribing width.

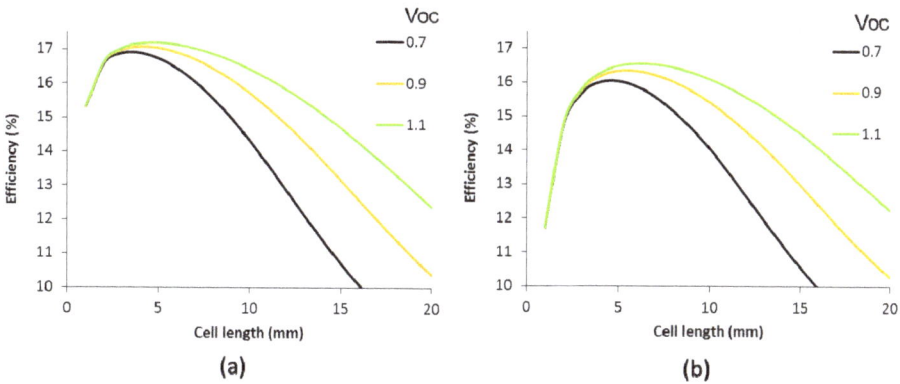

Figure 5. Efficiency of solar panels as a function of the individual cell length for different open circuit voltages (Voc in V) for a scribe width of 150 µm (**a**) and 350 µm (**b**). The front contact consists of a TCO of 10 Ω/sq.

If the scribing width is increased to 350 µm, the optimal cell length increases and hence the impact of the Voc on the maximum efficiency, as shown in Figure 5b. In other words, high Voc cells are less sensitive to scribing width than cells with a low Voc. Therefore, the absorber material not only has an impact on the maximum cell efficiency, but also on the cell layout.

2.3. Cells with Metallic Grid

For cells with a metallic grid on top of the TCO, it was found that a TCO of 50 Ω/sq is preferable over a large range of finger widths [16]. Therefore, Figure 6 shows the efficiency as a function of the cell length for cells with a 50 Ω/sq TCO supplied with a metallic finger grid with various finger heights (H_F). We also show the values for cells with a 10 Ohm/sq TCO front contact (black line).

For a scribing with of 150 μm (see Figure 6a), the efficiency increases from just below 17% to 17.8% when a high finger grid is used. For lower finger grid, the efficiency is somewhat lower and the cell length is also smaller. Nevertheless, even for a finger height of 1 μm, the increase in efficiency is 0.5 absolute %. This gain increases when a wider scribing area of 350 μm is taken into account. This is logical, because a TCO only approach cannot accommodate as long cells as compared to TCO supplemented with a finger grid, which show optimal cell lengths that are about twice that of the TCO only configuration. Therefore, the scribe area forms a lower proportion of the total area for longer cells and scribe related losses are proportionally reduced.

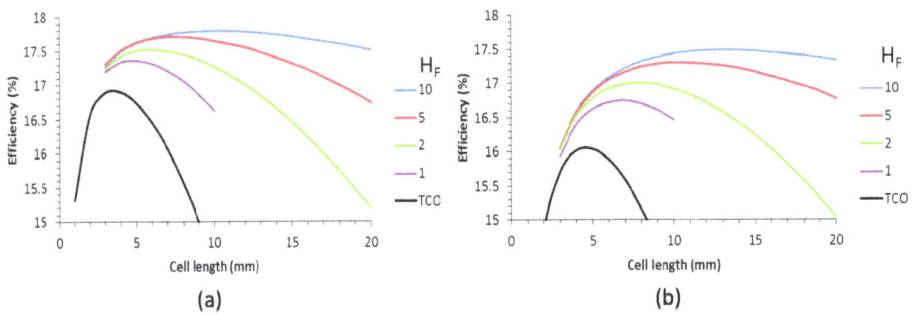

Figure 6. Efficiency of solar panels as function of the individual cell length for TCO-plus-grid front contact with different finger heights (H_F, in μm) for a scribe width of 150 μm (**a**) and 350 μm (**b**). The cell was based on a Voc of 0.7 V and the finger width is 20 μm. The TCO in the legend refers to calculations with a cell with a TCO of 10 Ω/sq.

A grid finger height of 10 μm could be hard to accomplish for printed lines and the data also indicate the impact of lower finger heights on the cell efficiency and the optimal cell length. On the other hand, the conductivity of the finger material used for this calculation is only 1/5 of the bulk conductivity of copper. Hence, finger material improvement can further increase the efficiency [25].

At present, a finger width of 20 μm is not compatible with large area printing technology. For this reason, wider fingers were also used for the calculations to assess the impact of finger width. Figure 7 shows the efficiency for cell lengths up to 20 mm, various finger heights for two different finger widths of 60 μm (Figure 7a,b) and 100 μm (Figure 7c,d). A comparison between a scribing width of 150 μm (Figure 7a,c) and 350 μm (Figure 7b,d) are also displayed.

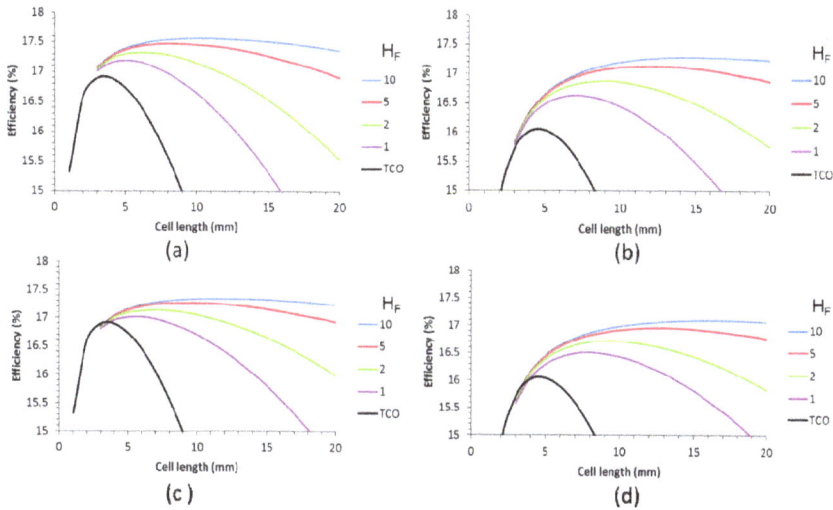

Figure 7. Efficiency of solar panels as a function of the individual cell length for different grid finger heights (H$_F$, in μm) for a scribe width of 150 μm (**a,c**) and 350 μm (**b,d**). The finger width is 60 μm (**a,b**) and 100 μm (**c,d**). The data are based on a Voc of 0.7 V.

Using a wider finger width than 20 μm decreases the efficiency benefit over the TCO only case. Nevertheless, for the presently available scribing width of 350 μm, the impact is still considerable and worth the additional manufacturing step. However, reducing the scribe width to 150 μm reduces the benefit of metallic grids.

2.4. Effect of Contact Resistance

One of the topics in thin film solar cells is the effect of contact resistance, although it is seldom mentioned [26,27]. The Mo/CIGS specific contact resistance was reported to be in the order of 0.08 Ohm cm^2 [28]. However, the specific contact resistance between TCO and Mo was found to be three orders of magnitude lower, in the range of 10^{-5} Ω cm^2 [29]. From the specific contact resistance (R$_{SCR}$), the contribution of the contact resistance to the overall resistance in the cell can be estimated. We calculated the contact resistance for a 1 cm^2 cell. This was done for different widths of overlap between the TCO and the Mo, as shown in the TCO/Mo contact width in Figure 8a. For a 1 cm^2 solar cell, typical total series resistances are between 1 and 2 ohm. For two specific contact resistances (R$_{scr}$), the contact resistance was calculated to be less than 0.02 Ω. As this is much lower than the typical series resistance in the cell, the impact is negligible.

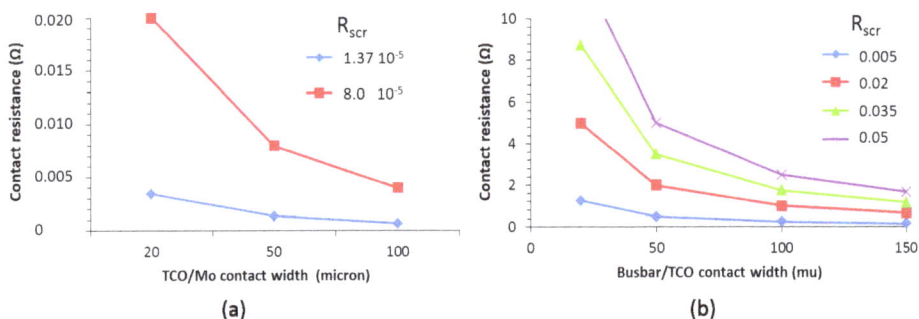

Figure 8. Contact resistance for a 1 cm^2 cell as a function of the width of the contact area between the TCO and the Mo (**a**) and the metal busbar and the TCO (**b**). The legend shows the specific contact resistance (R_{scr}, in Ω cm^2). The cell length is 5 mm.

For printed lines, the specific contact resistance between the metal and the TCO is between 0.01 and 0.05 Ω cm^2. At present, the factors underlying the contact resistance is under research in our group and include the ink material, the ink curing conditions and the TCO sheet resistance. The contribution of the contact resistance was calculated for a variety of various busbar/TCO overlap (contact) widths and specific contact resistances. Clearly, Figure 8b shows that the contact resistance is much higher than for the TCO/Mo case and can give a significant contribution to the overall series resistance.

For the case of the metallic interconnect combined with the metallic finger grid, the contact resistance was also calculated. In this case, the busbar width was taken to be 50 µm and the finger widths of 20, 60 and 100 µm were used with a finger spacing is 0.7, 1.6 and 2.1 mm, respectively. As a result of the higher contact area between the metal and the TCO, the contact resistance drops to values below 1.5 Ohm, as shown in Figure 9. This makes the system more robust against the occurrence of contact resistance between the printed metal and the TCO, although it is not negligible.

The effect of the contact resistance on the cell performance was calculated for the case without and with metallic grid. For the case without metallic grid, the data are presented in Figure 10 for a scribe with of 150 µm and 350 µm. We have used the range of specific contact resistance between 0.01 and 0.1 Ω cm^2. The black lines indicate the case without contact resistance (TCO interconnect). A specific contact resistance of 0.01 Ω cm^2 has only minimal impact on the cell efficiency. However, for higher specific contact resistances, the impact is larger and the efficiency drops several absolute percent for the highest specific contact resistances calculated. For a scribe width of 350 μm, the drop in efficiency is even more dramatic. This is caused by the fact that the wider scribing width induces a higher optimal cell length. The longer cells generate more current and translate into a larger effect of the series resistance. In this respect, the occurrence of contact resistance is an extra motivation to minimize the scribe width.

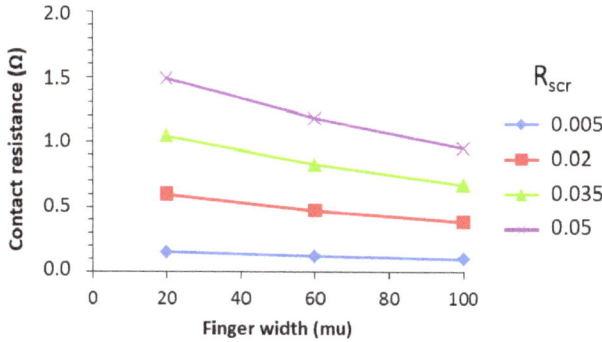

Figure 9. Contact resistance for a cell of 1 cm^2 for a cell with metal interconnect and fingers as a function of the finger width for various specific contact resistances (R_{scr}, in Ω cm^2). Ω 2.5. Impact of Contact Resistance on Cell Performance.

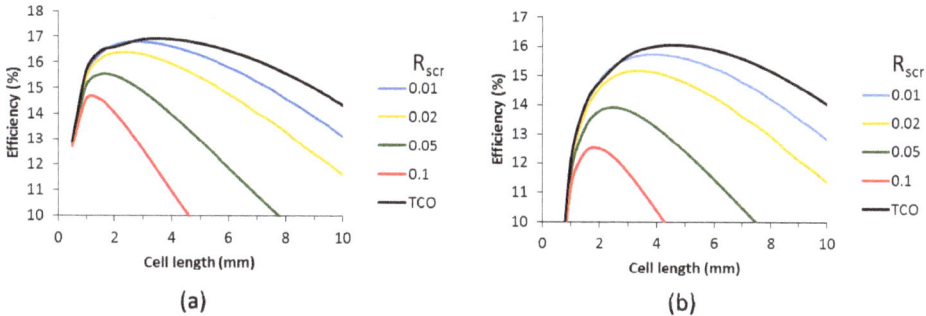

Figure 10. Efficiency as a function of the cell length and specific contact resistance for cells with 150 μm (**a**) and 350 μm (**b**) scribe width.

For cells with a 50 Ω/sq TCO supplemented with a metallic finger grid, the impact of the specific contact resistance was calculated for finger widths of 20 μm and 60 μm and various finger heights, as shown in Figure 11. A scribing width of 150 μm was used. We have included lower specific contact resistances to demonstrate that extremely low values do not impact the cell efficiency. However, from a specific contact resistance of 0.01 and upward, a consistent decrease in cell efficiency and optimal cell length is seen. Above a R_{scr} of 0.02, the efficiency enhancement by the metallic grid compared to the TCO is only very small. Higher finger grids can

123

compensate for this to some extent, but nevertheless, Figure 11 indicates that for a competitive performance of finger grids and metallic interconnect over the classic TCO interconnect, the R_{scr} should be at least 0.02 Ω cm^2.

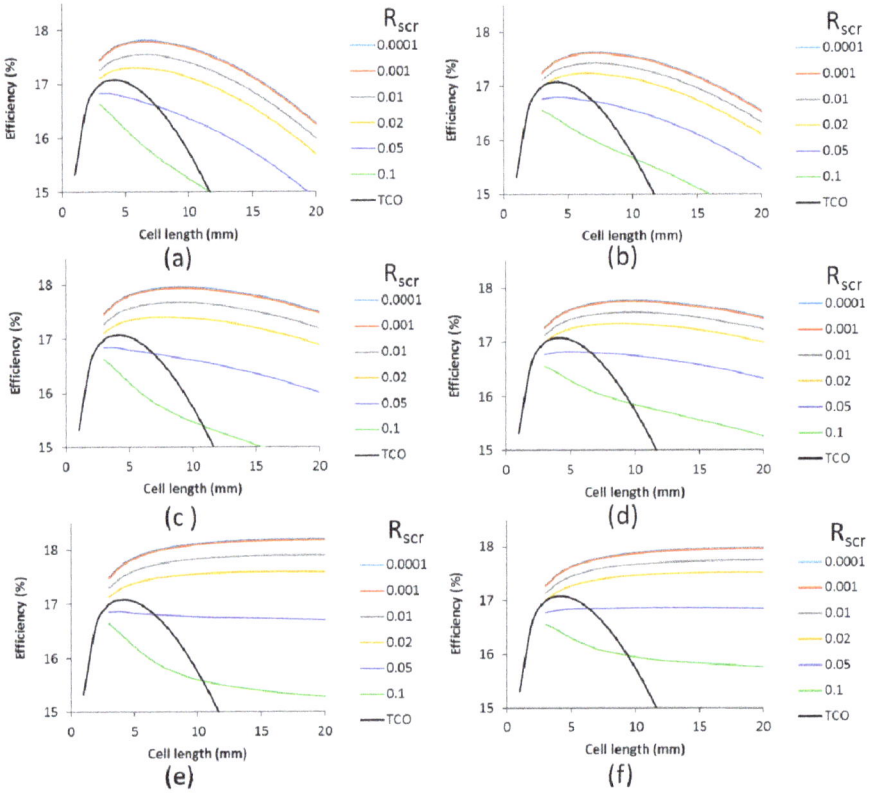

Figure 11. Efficiency as a function of the cell length for various specific contact resistances (see legend, R_{scr}, in Ω cm^2), and TCO (without contact resistance) for cells with finger grid width of 20 µm (**a,c,e**) and 60 µm (**b,d,f**) and a height of 2 µm (**a,b**), 5 µm (**c,d**) and 50 µm (**e,f**). Calculations were based on a 19% small cell.

The effect of the R_{scr} is smaller for wider grids. This can be explained by the larger contact area between the metal and the finger. As a result, for an R_{scr} of 0.02 Ohm cm^2 there is little difference in efficiency between the 20 µm and the 60 µm grid widths. The benefit of the lower shadow of the narrower grid finger is compensated by the higher contact resistance loss. This is independent of the grid height. Obviously, for higher grid fingers, the range of the applicable cell length increases, but the impact of the R_{scr} is similar. A longer cell increases both the contact area and the generated current and these two factors counterbalance each other.

In contrast, Figure 10 shows an increased impact of R_{scr} with cell length, as in this case, the longer cell length increases the current density, but the TCO metal contact area (busbar only) remains the same. For all cells with a metallic interconnect, the cells with a metallic grid show a higher cell efficiency (Figure 11) compared to the cells with only a TCO as the front contact for similar R_{scr} (Figure 10).

2.5. Influence of Illumination Power

Solar panels and solar cells are tested and certified at an illumination power of 1000 W/m^2 (also denoted as one sun). Therefore, the panel configuration is usually optimized for this high intensity. However, in northwest Europe, this high power is seldom reached. In real life, much of the power generated by solar panels is actually around an illumination power of 500 W/m^2. For cells without a metallic grid, the influence of the illumination power was investigated with variation of the cell length, as shown in Figure 12a. Seemingly, as the illumination power decreases, the impact of the cell length drops. However, when these data are normalized, as shown in Figure 12b, it is seen that the relative power is merely shifted toward somewhat higher cell lengths and the impact is reduced for longer cells. Nevertheless, down to an illumination power of 0.5 suns, the cell length remains a critical part of the configuration optimization.

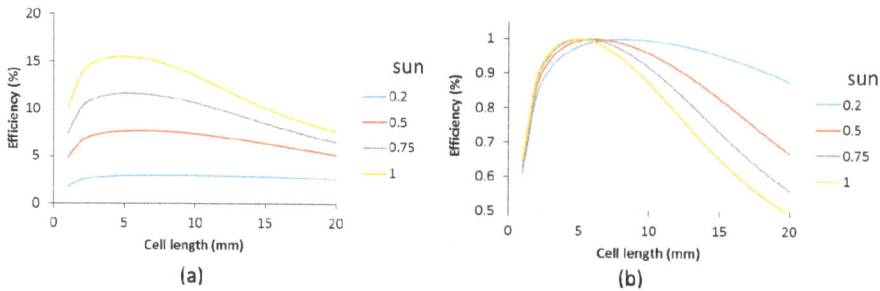

Figure 12. Efficiency as function of the cell length for different light intensities (see legend in sun units, whereby one sun is 1000 W/m^2): (**a**) calculated values; (**b**) normalized values.

For cells with a finger grid, the cell efficiency seems to become less affected by the cell length, as shown in Figure 13, which shows efficiency as function of the cell length for illumination powers from 0.2 to 1 sun in Figure 13a–d. Note that for each graph, the minimum value on the x-axis is about half of the maximum value to facilitate comparison with Figure 12b.

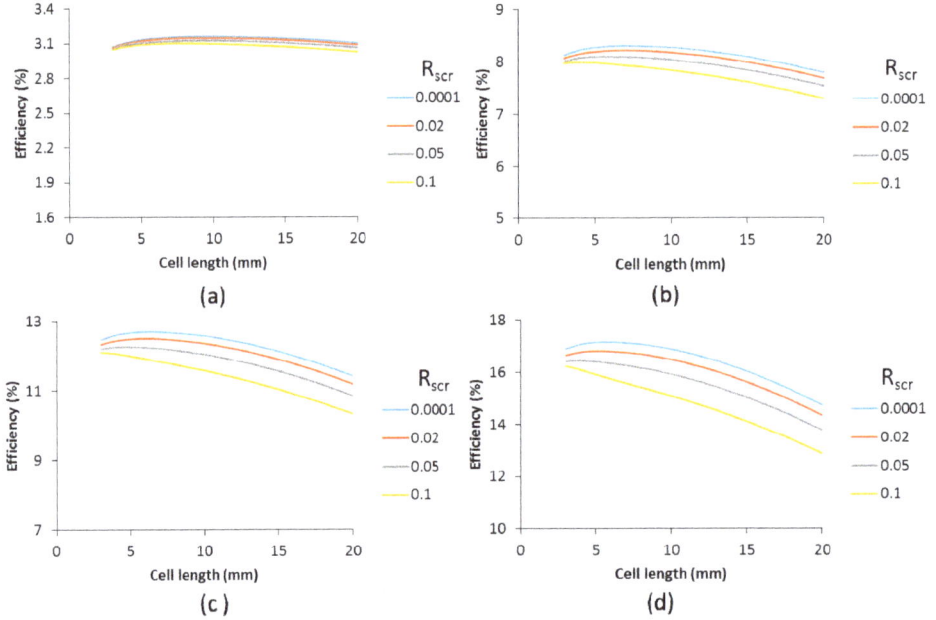

Figure 13. Efficiency as function of the cell length for various specific contact resistances (R_{scr}, in Ohm cm^2) for light intensity of: 0.2 sun (**a**); 0.5 sun (**b**); 0.75 sun (**c**); and 1 sun (**d**).

An additional observation is that lower illumination intensity reduces the impact of the specific contact resistance. In other words, the expected impact of the specific contact resistance for cells with metallic grids on the yearly yield of a solar panel is less than posed in Figure 10, which was based on an illumination power of one sun. Although this suggests that the R_{scr} is a less severe bottleneck, an R_{scr} below 0.02 Ω cm^2 is still highly recommended.

3. Experimental Section

For the TCO only case, we have used five TCOs with various sheet resistances. The transmittance increases with sheet resistance, and this depends on the specific TCO and deposition method used. We have used the data shown in Figure 3 to represent the additional optical loss by the TCO and the values are aimed to reflect industrial application methods.

For the case of TCO plus metallic grid, a wide range of TCO values were calculated and we present here only data for a TCO sheet resistance of 50 Ω/sq. This work only discusses a finger grid configuration, because in previous work, it was determined that this is the most effective grid design for monolithically integrated thin film PV [24]. The material characteristics of the metal in the model consisted of a conductivity of 1/5 of the bulk conductivity of copper to represent a moderate quality conductive ink [30]. The specific contact resistance was varied between 0.0001 and 0.1.

The modeling was performed in Comsol and the data weres further processed in Matlab. As input, the single diode description of a solar cell with a maximum efficiency of 19% was used with the following equation: $J = A - B * (e^{-C*V} - 1)$, where J is the current density (A/m^2) and V is voltage (V). The constants A, B and C are given in Table 1. In the model, voltage between 0 and Voc give the curves presented in Figure 2a. The constants are chosen in such a way that the maximum efficiency is 19% and the fill factor is 80%. Table 1 also shows the voltage and current density (Vmpp and Impp, respectively) at which this 19% is obtained.

Table 1. Constants used for the IV curves on which the modeling was based.

Material	Voc (V)	A	B	C	Vmpp (V)	Impp (A/m^2)
CIGS	0.7	337.5	1.0×10^{-6}	28.1	0.6	318.5
CdTe	0.9	262.7	1.0×10^{-6}	21.55	0.77	248.1
Perovskite	1.1	215.7	1.5×10^{-6}	17.05	0.94	203.3

4. Conclusions

The impact of the front contact design and interconnection material options were calculated for thin film solar cells. This includes many factors and variation of the TCO sheet resistance, scribing area, cell length, finger dimensions, contact resistance and illumination power were assessed. Metallic grids have a benefit in terms of higher solar cell efficiency and this also enables longer cell lengths. However, contact resistance between the metal and the TCO material can reduce this benefit somewhat.

Author Contributions: Author Contributions: Joop van Deelen contributed to the concept, the modeling, supervised the research work and wrote the manuscript, Yasemin Tezsevin performed the contact resistance measurements; and Marco Barink contributed to the modeling.

Conflicts of Interest: Conflicts of Interest: The authors declare no conflict of interest.

References

1. Jeng, M.J.; Chen, Z.Y.; Xiao, Y.L.; Chang, L.B.; Ao, J.; Sun, Y.; Popko, E.; Jacak, W.; Chow, L. Improving Efficiency of Multicrystalline Silicon and CIGS Solar Cells by Incorporating Metal Nanoparticles. *Materials* **2015**, *8*, 6761–6771.
2. Jackson, P.; Hariskos, D.; Wuerz, R.; Kiowski, O.; Bauer, A.; Powalla, M.; Friedlmeier, T.M. Properties of Cu(In,Ga)Se$_2$ Solar Cells with New Record Efficiencies up to 21.7%. *Phys. Status Solidi R* **2015**, *9*, 28–31.
3. Green, M.A.; Emery, K.; Hishikawa, Y.; Warta, W.; Dunlop, E.D. Solar cell efficiency tables (version 46). *Prog. Pothovoltaics.* **2015**, *23*, 8025–812.

4. Vermang, B.; Wätjen, J.T.; Fjällström, V.; Rostvall, F.; Edoff, M.; Kotipalli, R.; Henry, F.; Flandre, D. Employing Si solar cell technology to increase efficiency of ultra-thin Cu(In,Ga)Se$_2$ solar cells. *Prog. Photovoltaics* **2014**, *22*, 1023–1029.

5. Vermang, B.; Wätjen, J.T.; Fjällström, V.; Rostvall, F.; Edoff, M.; Gunnarsson, R.; Pilch, I.; Helmersson, U.; Kotipalli, R.; Henry, F.; Flandre, D. Highly reflective rear surface passivation design for ultra-thin Cu(In,Ga)Se$_2$ solar cells. *Thin Solid Films* **2015**, *582*, 300–303.

6. Stadler, A. Transparent Conducting Oxides—An Up-To-Date Overview. *Materials* **2012**, *5*, 661–683.

7. Calnan, S. Applications of Oxide Coatings in Photovoltaic Devices. *Coatings* **2014**, *4*, 162–202.

8. Westin, P.O.; Zimmermann, U.; Ruth, M.; Edoff, M. Next generation interconnective laser patterning of CIGS thin film modules. *Sol. Energ. Mater. Sol. Cell* **2011**, *95*, 1062–1068.

9. Bartlome, R.; Strahm, B.; Sinquin, Y.; Feltrin, A.; Ballif, C. Laser applications in thin-film photovoltaics. *Appl. Phys. B* **2010**, *100*, 427–436.

10. Van Deelen, J.; Barink, M.; Rendering, H.; Voorthuizen, P.; Hovestad, A. Improvement of transparent conducting materials by metallic grids on transparent conductive oxides. *Thin Solid Films* **2014**, *555*, 159–162.

11. Hovestad, A.; Rendering, H.; Maijenburg, A.W. Patterned electrodeposition of interconnects using microcontact printing. *J. Appl. Eletrochem.* **2012**, *42*, 753–761.

12. Malm, U.; Edoff, M. Influence from front contact sheet resistance on extracted diode parameters in CIGS solar cells. *Prog. Photovolt.* **2008**, *16*, 113–121.

13. Brecl, K.; Topic, M. Simulation of losses in thin-film silicon modules for different configurations and front contacts. *Prog. Photovolt.* **2008**, *16*, 479–488.

14. Koishiyev, G.T.; Sites, J.R. Impact of sheet resistance on 2-D modeling of thin-film solar cells. *Sol. Energ. Mater. Sol. Cell* **2009**, *93*, 350–354.

15. Galagan, Y.; Coenen, E.W.C.; Sabik, S.; Gorter, H.H.; Barink, M.; Veenstra, S.C.; Kroon, J.M.; Andriessen, R.; Blom, P.W.M. Evaluation of ink-jet printed current collecting grids and busbars for ITO-free organic solar cells. *Sol. Energ. Mater. Sol. Cell* **2012**, *104*, 32–38.

16. Van Deelen, J.; Barink, M.; Klerk, L.; Voorthuijzen, W.P.; Hovestad, A. Efficiency loss prevention in monolithically integrated thin film solar cells by improved front contact. *Prog. Photovolt.* **2015**, *23*, 498–506.

17. Van Deelen, J.; Rendering, H.; het Mannetje, H.; Klerk, L.; Hovestad, A. Metallic Grids for Low Resistive Transparent Conductors: Modeling and Experiments. In Proceedings of the 38th IEEE Photovoltaic Specialists Conference, Austin, TX, USA, 3–8 June 2012.

18. Van Hest, M.F.A.M.; Curtis, C.J.; Miedaner, A.; Pasquarelli, R.M.; Kaydanova, T.; Hersh, P.; Ginley, D.S. Direct-write contacts: Metallization and contact formation. In Proceedings of the 33rd IEEE Photovoltaic Specialists Conference, San Diego, CA, USA, 11–16 May 2008.

19. Hersh, P.A.; Curtis, C.J.; Van Hest, M.F.A.M.; Kreuder, J.J.; Pasquarelli, R.; Miedaner, A.; Ginley, D.S. Inkjet printed metallization for Cu(In$_{1-x}$Ga$_x$)Se$_2$ photovoltaic cells. *Prog. Photovolt.* **2011**, *19*, 973–976.

20. Retterstol Olaisen, B.; Woldegiorgis, S.; Westin, P.O.; Edoff, M.; Stolt, L.; Holt, A.; Stensrud Marstein, E. CIGS minimodules with screen printed front contacts. In Proceedings of the 15th International Photovoltaic Science and Engineering Conference, Shanghai, China, 10–15 October 2005.

21. Wipliez, L.; Löffler, J.; Heina, M.C.R.; Slooff-Hoek, L.H.; De Keijzer, M.A.; Bosman, J.; Soppe, W.J.; Schoonderbeek, A.; Stutte, U.; Furthner, F.; *et al.* Monolithic series interconnection of flexible thin-film PV devices. In Proceedings of the 26th European Photovoltaic Solar Energy Conference, Hamburg, Germany, 5–9 September 2011.

22. Kamikawa-Shimizu, Y.; Komaki, H.; Yamada, A.; Ishizuka, S.; Iioka, M.; Higuchi, H.; Kakano, M.; Matsubara, K.; Shibata, H.; Niki, S. Highly efficient $Cu(In,Ga)Se_2$ thin film submodule fabricated using a three-stage process. *Appl. Phys. Express* **2013**, *6*, 112303.

23. Van Deelen, J.; Rendering, H.; Hovestad, A. Recent progress in transparent conducting materials by use of metallic grids on metaloxides. *Mater. Res. Symp. Proc.* **2011**, *1323*, 17–21.

24. Komaki, H.; Furue, S.; Yamada, A.; Ishizuka, S.; Shibata, H.; Matsubara, K.; Niki, S. High-efficiency CIGS submodules. *Prog. Photovolt.* **2012**, *20*, 595–599.

25. Van Deelen, J.; Klerk, L.; Barink, M. Optimized grid design for thin film solar panels. *Sol. Energ.* **2014**, *107*, 135–144.

26. Fields, J.D.; Dabney, M.S.; Bollinger, V.P.; Van Hest, M.F.A.M. Printed monolithic interconnects for photovoltaic applications. In Proceedings of the 40th IEEE Photovoltaic Specialists Conference, Denver, CO, USA, 8–13 June 2014.

27. Tait, J.G.; La Notte, L.; Melkonyan, D.; Gehlhaar, R.; Cheyns, D.; Reale, A.; Heremans, P. Electrical properties of patterned photoactive layers in organic photovoltaic modules. *Sol. Energ. Mater. Sol. C* **2016**, *144*, 493–499.

28. Assmann, L.; Bernede, J.C.; Drici, A.; Amory, C.; Halgand, E.; Morsli, M. Study of the Mo thin films and Mo/CIGS interface properties. *Appl. Surf. Sci.* **2005**, *246*, 159–166.

29. Oertel, M.; Götz, S.; Cieslak, J.; Haarstrich, J.; Metzner, H.; Wesch, W. Measurement of the zinc oxide-molybdenum specific contact resistance for applications in $Cu(In,Ga)Se_2$-technology. *Thin Solid Films* **2011**, *519*, 7545–7548.

30. Van Deelen, J.; Frijters, C.; Suijlen, M.; Barink, M. Influence of the grid and cell design on the power output of thin film $Cu(InGa)Se_2$ cells. *Thin Solid Films* **2015**, *594*, 225–231.

129

Heterojunctions of p-BiOI Nanosheets/n-TiO$_2$ Nanofibers: Preparation and Enhanced Visible-Light Photocatalytic Activity

Kexin Wang, Changlu Shao, Xinghua Li, Fujun Miao, Na Lu and Yichun Liu

Abstract: p-BiOI nanosheets/n-TiO$_2$ nanofibers (p-BiOI/n-TiO$_2$ NFs) have been facilely prepared via the electrospinning technique combining successive ionic layer adsorption and reaction (SILAR). Dense BiOI nanosheets with good crystalline and width about 500 nm were uniformly assembled on TiO$_2$ nanofibers at room temperature. The amount of the heterojunctions and the specific surface area were well controlled by adjusting the SILAR cycles. Due to the synergistic effect of p-n heterojunctions and high specific surface area, the obtained p-BiOI/n-TiO$_2$ NFs exhibited enhanced visible-light photocatalytic activity. Moreover, the p-BiOI/n-TiO$_2$ NFs heterojunctions could be easily recycled without decreasing the photocatalytic activity owing to their one-dimensional nanofibrous structure. Based on the above, the heterojunctions of p-BiOI/n-TiO$_2$ NFs may be promising visible-light-driven photocatalysts for converting solar energy to chemical energy in environment remediation.

Reprinted from *Materials*. Cite as: Wang, K.; Shao, C.; Li, X.; Miao, F.; Lu, N.; Liu, Y. Heterojunctions of p-BiOI Nanosheets/n-TiO$_2$ Nanofibers: Preparation and Enhanced Visible-Light Photocatalytic Activity. *Materials* **2016**, *9*, 90.

1. Introduction

Semiconductor photocatalysts have been promising media to degrade pollutants through converting solar energy to chemical energy [1]. In recent years, bismuth oxyhalides BiOX (X = Cl, Br, I), as p-type semiconductors, have attracted considerable attentions owing to their unique layered crystal structure and indirect optical transition characteristic. The layer structure consists of (Bi$_2$O$_2$)$^{2+}$ layers interleaved by double halogen atoms layers, which can present an internal static electric field and induce the separation of photogenerated carries more efficiently. Furthermore, the indirect transition band gap is necessary for electrons to be emitted to valence band by a certain k-space distance, which reduces the recombination probability of photoelectrons and holes [2,3]. These advantages lead to remarkable photocatalytic activities in degradation of organic compounds [4–7]. Notably, the p type BiOI (p-BiOI) nanostructures have the narrowest band gap (~1.8 eV) among BiOX (X = Cl, Br, I) nanostructures. Therefore, the p-BiOI nanostructures are considered excellent visible light photocatalysts that can utilize more solar light energy in photocatalysis. Nowadays, many efforts have been employed to improve the photocatalytic efficiency of p-BiOI nanostructures [8,9]. Constructing p-n heterojunctions is considered an effective method to improve the separation efficiency of photogenerated carries due to their strong internal electric field [10–13]. Many kinds of p-n heterojunctions based on p-BiOI, such as $BiOI/ZnTiO_3$ [14], $BiOI/Zn_2SnO_4$ [15], $BiOI/ZnO$ [16], $BiOI/Bi_4Ti_3O_{12}$ [17], *etc.*, have been reported with increased photocatalytic activity.

Among many n type semiconductors, Titanium dioxide (TiO_2) nanostructures have been widely studied as good photocatalysts due to their high efficiency, chemical stability, nontoxicity, low cost, *etc.* [18–24]. Coupling p-BiOI nanostructures with n-TiO_2 nanostructures to form p-BiOI/n-TiO_2 heterojunctions would hinder the recombination of photogenerated carries more effectively. To date, p-BiOI/n-TiO_2 nanoparticles have been widely reported with enhanced visible-light photocatalytic activity [25–27]. However, the suspended nanoparticles tend to aggregate during the synthesis process and be lost in the separation and recycling process, resulting in a reduction of specific surface area and photocatalytic performance. Compared with nanoparticles, one-dimensional nanofibers with a high surface-to-volume ratio are more favorable for both photocatalytic activity and recycling characteristics [28,29]. In fact, our group has previously constructed heterojunctions of p-BiOCl nanosheets/n-TiO_2 nanofibers [30] and p-MoO_3 nanosheets/n-TiO_2 nanofibers [31], both of which show enhanced ultraviolet photocatalytic activities and recycling properties. Therefore, there is interest in constructing p-BiOI/n-TiO_2 heterojunctions using electrospun TiO_2 nanofibers as n type semiconductor because of the following advantages: (1) besides the internal electric field of the p-n heterojunction, the one-dimensional characters of TiO_2 nanofibers could act as charge transfer channels facilitating higher charge separation efficiencies; (2) the three-dimensional open structure and large specific surface area of TiO_2 nanofibers provide more active sites for the assembly of secondary nanostructures with high densities; and (3) their nanofibrous nonwoven web structure can be easily separated from fluid by sedimentation.

Taking the above factors into account, in this work, the p-type BiOI nanosheets were successfully synthesized on n-type electrospun TiO_2 nanofibers by successive ionic layer adsorption and reaction (SILAR) at room temperature. The contents of BiOI in the heterojunctions of p-BiOI nanosheets/n-TiO_2 nanofibers (p-BiOI/n-TiO_2 NFs) could be well controlled by adjusting the cycles of SILAR. X-ray photoelectron spectra showed that both Ti 2p peaks of p-BiOI/n-TiO_2 NFs shifted to higher binding energies than that of TiO_2 nanofibers, suggesting effective electrons transfer from TiO_2 to BiOI in the formation of p-n heterojunction. The p-BiOI/n-TiO_2 NFs exhibited favorable visible-light photocatalytic activity for degradation of methyl orange (MO), which can be ascribed to the high specific surface area and the as-formed p-n heterojunctions. Moreover, the heterojunctions of p-BiOI/n-TiO$_2$ NFs could be well recycled by sedimentation without decreasing their photocatalytic activities.

2. Results and Discussion

2.1. Morphologies

Figure 1 shows the scanning electron microscopy (SEM) images of different samples. It can be seen that TiO$_2$ nanofibers with a diameter about 250 nm are relative smooth without secondary structures. After SILAR process, as observed in Figure 1b–d, the TiO$_2$ nanofibers become rough and are decorated with ultrathin BiOI nanosheets and remain as one-dimension nanofibers structure. When the cycles of SILAR process increase to 30 for the BiOI/TiO$_2$-C30, the amount of the BiOI nanosheets increases significantly compared with those of BiOI/TiO$_2$-C10 and BiOI/TiO$_2$-C20. These results suggest that the heterojunctions of p-BiOI/n-TiO$_2$ NFs may have a higher specific surface area, which is good for the photocatalytic reactions.

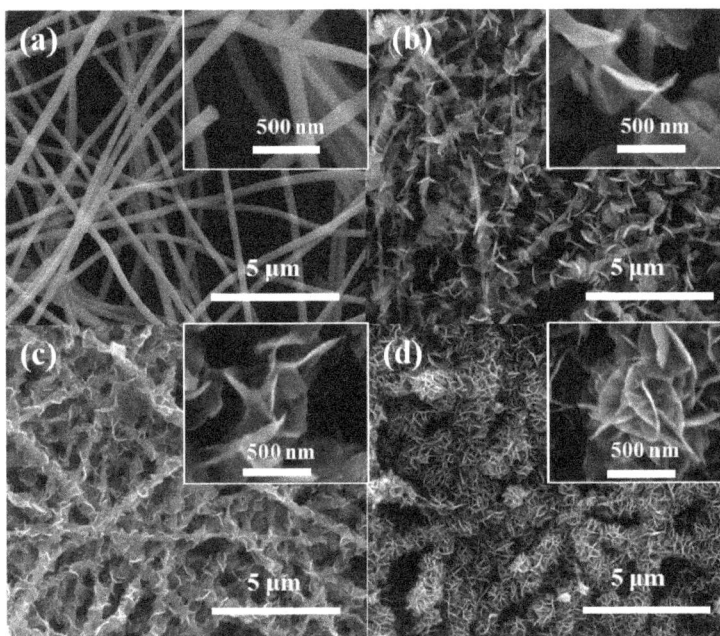

Figure 1. (**a**) SEM images of TiO$_2$ nanofibers; (**b**) BiOI/TiO$_2$-C10; (**c**) BiOI/TiO$_2$-C20; and (**d**) BiOI/TiO$_2$-C30 at low magnification and high magnification (insets).

Figure 2a,b shows the typical TEM images of BiOI/TiO$_2$-C30. It can be observed that numerous BiOI nanosheets are randomly distributed on TiO$_2$ nanofibers. The BiOI nanosheets are very thin, which coincides with the results of SEM observations.

The high-resolution transmission electron microscopy (HRTEM) image of the heterojunctions displays two types of lattice fringes, as shown in Figure 2b. One set of the fringes spacing is *ca.* 0.35 nm, corresponding to the (101) plane of the anatase crystal structure of TiO_2. Another set of the fringes spacing measures *ca.* 0.28 nm, which corresponds to the (110) lattice spacing of the BiOI. It indicates that heterojunctions are composed of TiO_2 nanofibers and BiOI nanosheets with exposed {001} facets. The exposed {001} facets may have excellent photocatalytic activity for BiOI under visible-light irradiation as reported [32]. Therefore, the surface reactivity may also be improved by decorating TiO_2 nanofibers with BiOI nanosheets.

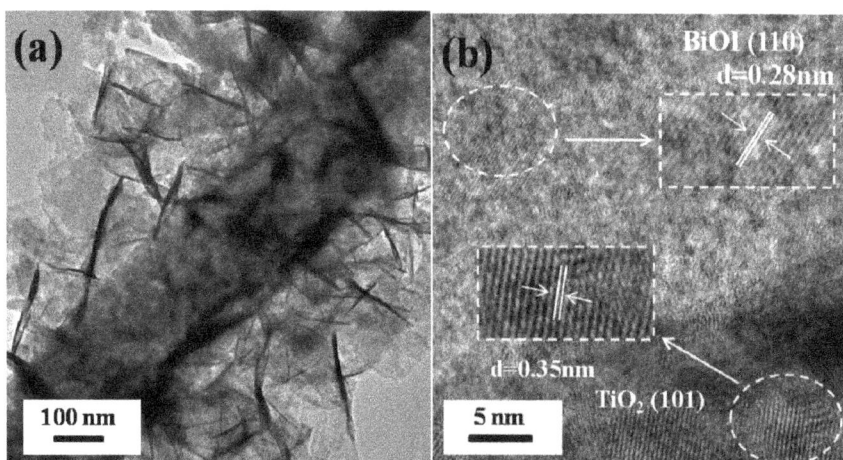

Figure 2. (a) TEM; and (b) HRTEM images of $BiOI/TiO_2$-C30.

2.2. Structure Characterization

Figure 3 shows the X-ray diffraction (XRD) patterns of pure TiO_2 nanofibers, p-BiOI/n-TiO_2 NFs and BiOI nanosheets. For TiO_2 nanofibers, all peaks are attributed to the anatase of TiO_2 (JCPDS No. 21-1272) and the rutile of TiO_2 (JCPDS No. 21-1276). For p-BiOI/n-TiO_2 NFs, besides the characteristic peaks of TiO_2 (solid and hollow diamonds), there are some new strong patterns that can be indexed as tetragonal phase of BiOI (JCPDS No. 73-2062). The diffraction peaks of BiOI (solid circles) are gradually intensified as the SILAR cycles increased from 0 to 30, as shown in Figure 3. No other characteristic peaks of impurities are observed. In particular, the domination of (110) plane in the pattern suggests that the exposed facets of BiOI nanosheets are mainly {001}. This result is consistent with SEM and TEM analyses.

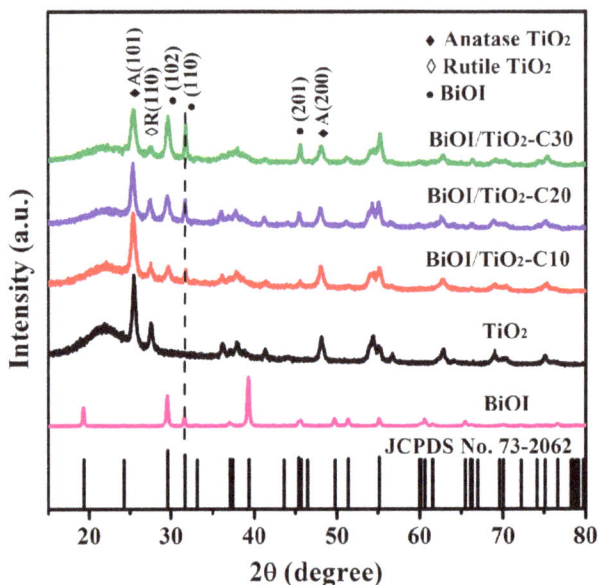

Figure 3. XRD patterns of different samples.

2.3. Composition and Chemical States

X-ray photoelectron spectroscopy (XPS) measurements are also performed to further investigate the chemical compositions and chemical states of p-BiOI/n-TiO$_2$ NFs. The typical high resolution XPS spectrum of Bi 4f is shown in Figure 4a. The peaks at 158.82 eV and 164.14 eV correspond to Bi 4f$_{3/2}$ and Bi 4f$_{1/2}$, respectively, which indicates a normal state of Bi^{3+} in BiOI/TiO$_2$-C30 [33]. Figure 4b reveals the high-resolution XPS spectrum of I 3d. The two peaks at 630.3 eV and 618.8 eV are attributed to I 3d$_{3/2}$ and I 3d$_{5/2}$, respectively, which indicates that the chemical state of iodine is I^{-1} in BiOI/TiO$_2$-C30 [34]. The deconvolution of the O 1s spectrum in Figure 4c implies that more than one chemical state of O 1s exists in the BiOI/TiO$_2$-C30. The peaks with lower binding energies at 529.8 eV and 531.5 eV correspond to the stronger Bi-O and Ti-O bond, respectively. The higher bonding energy of 532.9 eV might be caused by adsorbed water and surface hydroxyl groups (O$_{OH}$), which may also lead to an enhanced photocatalytic property [35]. The splitting between Ti 2p$_{1/2}$ and Ti 2p$_{3/2}$ are both 5.7 eV for TiO$_2$ and BiOI/TiO$_2$-C30, suggesting a normal state of Ti^{4+} in pure TiO$_2$ nanofibers and BiOI/TiO$_2$-C30 [36,37]. However, for BiOI/TiO$_2$-C30, the binding energy of Ti 2p$_{3/2}$ locates at 458.7 eV, which is about 0.4 eV higher than that of pure TiO$_2$ nanofibers (458.3 eV). This can be explained as follow: when p type BiOI nanosheets are deposited on n type TiO$_2$ nanofibers, the electrons in TiO$_2$ nanofibers would diffuse to BiOI, forming p-n heterojunctions; thus, in the space charge region, TiO$_2$ is positively charged which could increase the binding energy of electrons in Ti 2p chemical states. Similar results have been observed in the heterojunctions of p-MoO$_3$ nanosheets/n-TiO$_2$ nanofibers [26].

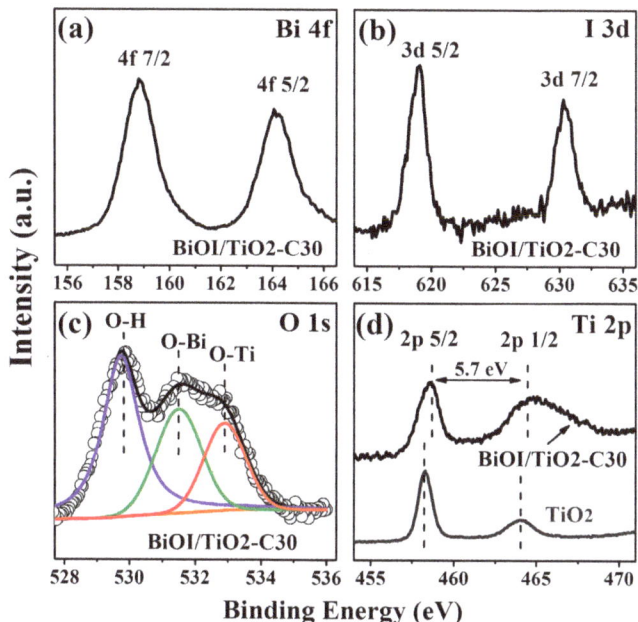

Figure 4. (a) XPS spectra of Bi 4f; (b) I 3d; and (c) O 1s for BiOI/TiO$_2$-C30; (d) XPS spectra of Ti 2p for TiO$_2$ nanofibers (bottom) and BiOI/TiO$_2$-C30 (top).

2.4. Nitrogen Adsorption

All the samples have typical type-IV N$_2$ adsorption-desorption isotherms with H1 hysteresis indicative of mesoporous structure (Figure 5). The curve of pure TiO$_2$ nanofibers implies a meso- and macropore structure. As we know, the precursor nanofibers of electrospun TiO$_2$ nanofibers consist of polymer and metal salt. During the calcination process, the decomposition of the polymer and metal salt can result in abundant hierarchical pores with a wide pore size distribution of more than 2 nm, as shown in Figure 5 inset. For BiOI/TiO$_2$-C10, there is an obvious hysteresis loop in the large relative pressure range of 0.9–1.0 (P/P$_0$), indicating the relatively large pore structure arising from the voids among the BiOI nanosheets on TiO$_2$ nanofibers. The specific surface areas of these samples are shown in Table 1. It is also worth noting that the BiOI/TiO$_2$-C10 exhibit lower specific surface areas than that of TiO$_2$ nanofibers, which can be ascribed to the deposition of BiOI nanosheets blocking the original pores on TiO$_2$. It can be clearly seen that the relative small pores (2–11 nm) are disappeared, which can be demonstrated by the pore size distribution of BiOI/TiO$_2$-C10 in Figure 5 inset. Compared to BiOI/TiO$_2$-C10, there is an obviously increased adsorption at high pressure with increased deposition of BiOI nanosheets on TiO_2 nanofibers for BiOI/TiO_2-C20 and BiOI/TiO_2-C30, along with increased specific surface areas, indicating the more and more abundant porosity structures. It is accepted that the porosity is relative to the amount of BiOI nanosheets depositing on TiO_2 nanofibers. Hence, the close arrangements of BiOI nanosheets on TiO_2 nanofibers (see SEM and TEM images) have resulted in the hierarchical porosity with wide pore size distributions, which are further confirmed by the corresponding pore size distributions in the inset of Figure 5. These results suggest that the BiOI/TiO_2 nanofibers with abundant porosity and large specific surface areas will increase the assessable surface areas of the catalyst with dye solution to achieve good photocatalytic activity. Particularly, the large amount of BiOI nanosheets depositing on TiO_2 nanofibers without independent nucleation will benefit the formation of more p-n heterojunctions as well as rapid charge transfer during the photocatalysis.

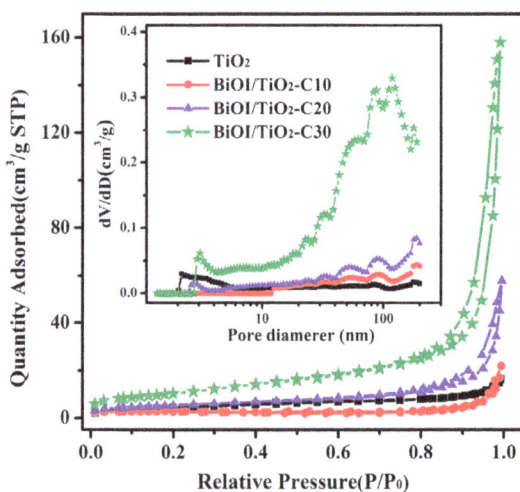

Figure 5. Typical N_2 gas adsorption desorption isotherms of different samples and their corresponding pore-size distributions (inset).

Table 1. SILAR cycles, BET specific surface area and photocatalysis reaction rates of different samples.

Samples	Cycles	BET Specific Surface Area (m²/g)	K_{app} (h^{-1})
TiO_2 NFs	0	15.88	0.000 ± 0.000
BiOI/TiO_2-C10	10	7.27	0.141 ± 0.002
BiOI/TiO_2-C20	20	19.19	0.197 ± 0.009
BiOI/TiO_2-C30	30	38.44	0.724 ± 0.095
M-BT (Bi:Ti = 0.4:1)	-	-	0.267 ± 0.024

2.5. Optical Properties

Figure 6 shows the UV-vis absorption spectra of TiO_2, $BiOI/TiO_2$-C10, $BiOI/TiO_2$-C20, $BiOI/TiO_2$-C30 and BiOI converted from corresponding diffuse reflectance spectra by means of the Kubelka–Munk function [28]:

$$F(R) = (1 - R)^2/2R = \alpha/S \qquad (1)$$

$$R = R_{Sample}/R_{BaSO4} \qquad (2)$$

where R, α, and S are the reflectance, absorption coefficient and scattering coefficient, respectively. It can be seen that TiO_2 exhibited a typical absorption characteristic of the wide band gap semiconductor with an edge about 380 nm, while pure BiOI with a strong absorption at about 630 nm in the visible light region, indicates that it is a narrow band gap semiconductor according to the equation $E_g = 1240/\lambda$, where E_g is the band gap (eV) and λ (nm) is the wavelength of the absorption edge in the spectrum. The band gap of TiO_2 and BiOI are estimated to be 3.2 eV and 1.9 eV, respectively. It is noted that the absorption edge of p-BiOI/n-TiO_2 NFs show significant red-shift from 393 to 500 nm with the increased amount of BiOI in the composite nanofibers. Based on the above, the increased amount of BiOI in p-BiOI/n-TiO_2 NFs extends light absorbing range, which is the precondition of effective photocatalytic activity.

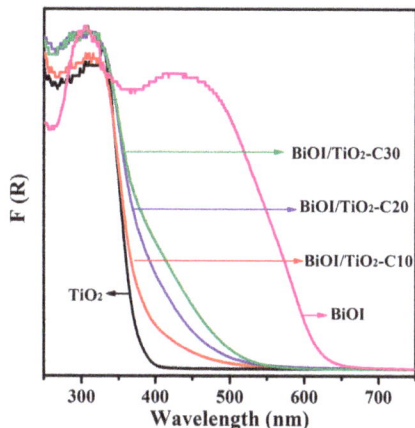

Figure 6. UV-vis absorption spectra of different samples.

2.6. Photocatalytic Properties

Figure 7a shows the photocatalytic activities of TiO_2 NFs, $BiOI/TiO_2$-C10, $BiOI/TiO_2$-C20, $BiOI/TiO_2$-C30 and the mechanical mixture of BiOI and TiO_2 (M-BT, the molar ratio of Bi:Ti = 0.4:1 based on energy dispersive X-ray (EDX) analysis in

Figure S1) on the degradation of methyl orange (MO) under visible-light irradiation (⩾420 nm). Before irradiation, the adsorption-desorption equilibrium of MO in the dark is established within 30 min over different samples. The time-dependent absorbance spectra of different samples are shown in Figure S1. The adsorption of BiOI/TiO$_2$-C30 increases significantly compared to other samples, which might be attributed to the high specific surface area. After 3 h irradiation, the photodegradation efficiencies of MO for BiOI/TiO$_2$-C30 are about 92%, in comparison to 60%, 66%, 38% and almost none for M-BT, BiOI/TiO$_2$-C20, BiOI/TiO$_2$-C10 and TiO$_2$ nanofibers, respectively. In Figure 7b, the kinetic linear fitting curves over different photocatalysts show that the photocatalytic degradation of MO followed a Langmuir-Hinshelwood apparent first-order kinetics model:

$$\ln C/C_0 = -kKt = -k_{\text{app}}t \qquad (3)$$

where C_0 is the initial concentration (mg/L) of the reactant; C is the concentration (mg/L); t is the visible-light irradiation time; k is the reaction rate constant (mg/(L·min)); and K is the adsorption coefficient of the reactant (L/mg); k_{app} is the apparent first-order rate constant (min^{-1}). The k_{app} of different samples are shown in Table 1. It is indicated that the photocatalytic activities is in the order of BiOI/TiO$_2$-C30 > BiOI/TiO$_2$-C20 > M-BT > BiOI/TiO$_2$-C10 > TiO$_2$. The above illuminates that the construction of p-n heterojunctions can effectively enhance the photocatalytic properties. Furthermore, the increased of the specific surface area and the amount of p-n heterojuctions obviously enhance the photocatalytic activity. Furthermore, the photocatalysis under UV-light irradiation (Figure S2) also demonstrates the above point.

To understand the photocatalytic properties of p-BiOI/n-TiO$_2$ NFs, a schematic diagram is proposed (Scheme 1). When p-type BiOI contacts n-type TiO$_2$, the diffusion of electrons and holes create an inner electric field where a space-charge region is formed at the interfaces of p-n heterojunction. Under visible-light irradiation, the photogenerated electrons transfer from the conduction band of BiOI to that of TiO$_2$, while the photogenerated holes stay at the valence band of BiOI. The recombination of photogenerated charge carrier is inhibited greatly in the heterojunctions of p-BiOI/n-TiO$_2$ NFs. Thus, the photogenerated electrons and holes can effectively take part in the photodegradation of MO under visible light. On the other hand, the nanofiber structures of TiO$_2$ can prevent the agglomeration of BiOI nanosheets and facilitate the transfer of the dye molecules during photocatalytic process. Moreover, the exposed surface of BiOI is mainly {001} facet, which is very active for photocatalytic reactions under visible-light irradiation [32]. Thus, the nanosheet structure of BiOI might also improve the surface reaction rates and contribute to the photocatalysis. It should be noted that the p-BiOI/n-TiO$_2$ NFs can be easily separated from an aqueous suspension for reuse due to their one-dimensional nanofibrous morphology. As shown in Figure 8, the photodegradation of MO on the p-BiOI/n-TiO$_2$ NFs was reused three times. Each experiment was carried out under identical conditions. Clearly, the photocatalytic activity of p-BiOI/n-TiO$_2$ NFs remains almost unchanged after three-cycles, suggesting that the BiOI/TiO$_2$ NFs have good stability and recycling properties.

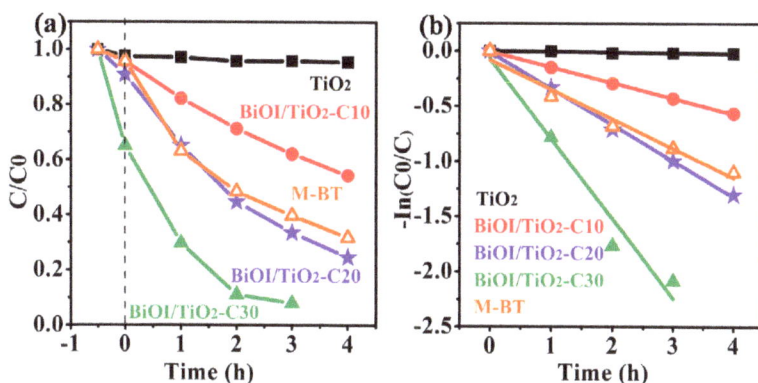

Figure 7. (a) Degradation curves of MO under visible light irradiation; and (b) the apparent first-order kinetics fitting over different samples.

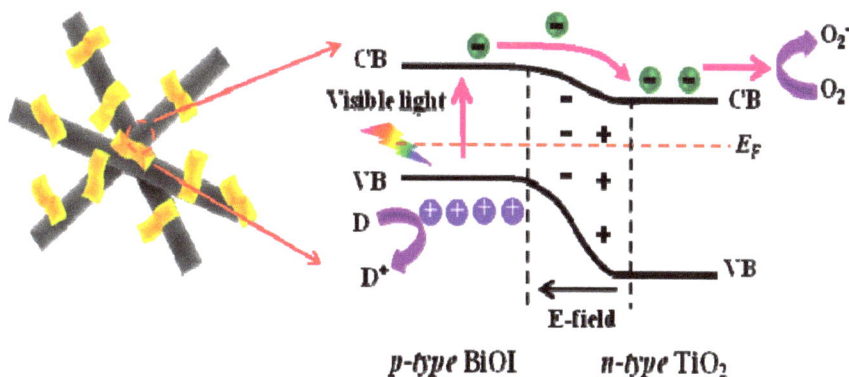

Scheme 1. Possible photocatalytic reactions of p-BiOI/n-TiO$_2$ NFs heterojunctions.

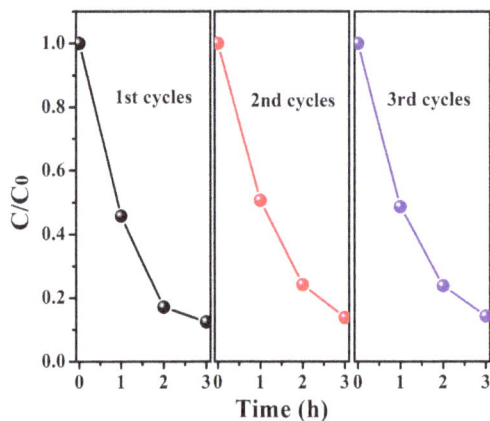

Figure 8. Photocatalysis tests of BiOI/TiO$_2$-C30 for three cycles.

3. Experimental

The synthesized process of samples is presented in Scheme 2.

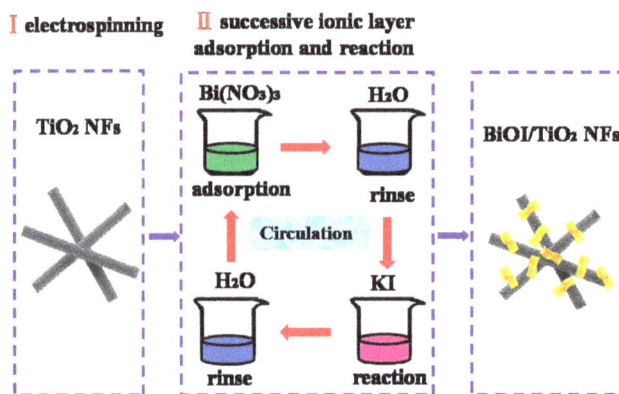

Scheme 2. Schematic illustration for the preparation of p-BiOI/n-TiO$_2$ NFs heterojunctions.

3.1. Fabrication of TiO$_2$ Nanofibers

Firstly, 1.6 g Poly(vinyl pyrrolidone) powder (PVP, Mw = 1,300,000) was added to a mixture of 20 mL absolute ethanol and 2 mL acetic acid in a Erlenmeyer flask. The obtained solution was stirred for 2 h to generate a homogeneous solution. Then, 2.0 mL Ti(OC$_4$H$_9$)$_4$ was added to the solution, the mixture was magnetically stirred for another 10 h at room temperature to make electrospinning precursor solution. Subsequently, the above precursor solutions were drawn into a hypodermic syringe

140

with a needle tip. Then, a high voltage source was connected to the needle tip while a sheet of aluminum foil was employed as the collector. The voltage between the needle tip and collector was set at 10 kV, and the distance was 15 cm. The as-collected nanofibers were calcined at a rate of 25 °C/h and remained for 2 h at 520 °C to obtain TiO_2 NFs.

3.2. Fabrication of $BiOI/TiO_2$ Nanofibers

The p-BiOI/n-TiO_2 NFs were synthesized through the SILAR process. Typically, 0.25 mM $Bi(NO_3)_3 \cdot 5H_2O$ solutions were prepared with deionized water as solution A, and equivalent concentration of KI solution were prepared as solution B. The TiO_2 nanofibers were first immersed into solution A for 2 min, rinsed with deionized water, and then immersed into solution B for 2 min, rinsing likewise. The four-step procedure forms one cycle and the BiOI would increase by repeating the cycles. A series of samples, with different cycles of 10, 20 and 30 were prepared and denoted as $BiOI/TiO_2$-C10, $BiOI/TiO_2$-C20 and $BiOI/TiO_2$-C30. After that, the samples were thoroughly rinsed with deionized water and allowed to dry at 60 °C overnight. All the samples are listed in Table 1. Pure BiOI nanosheets were prepared by mixing solution A and B, then rinsed and dried.

3.3. Characterizations

Sanning electron microscopy (SEM, Quanta 250 FEG, FEI, Hillsboro, OR, USA) and high-resolution transmission electron microscopy (HRTEM; JEOL JEM-2100, JEOL, Tokyo, Japan) were used to characterize the morphologies of the products. The X-ray diffraction (XRD) measurements were carried out using a D/max 2500 XRD spectrometer (Rigaku, Tokyo, Japan) with a Cu Kα line of 0.1541 nm. The X-ray photoelectron spectroscopy (XPS) was performed on a VG-ESCALAB LKII instrument (VG, Waltham, UK) with Mg KαADES (hυ = 1253.6 eV) source at a residual gas pressure of below 10^{-8} Pa. The specific surface area of the samples were measured with a Micromeritics ASAP 2010 instrument (Micromeritics, Norcross, GA, USA) and analyzed by the Brunauer-Emmett-Teller (BET) method. The UV-vis diffuse reflectance spectra were measured at room temperature with a UH4150 spectrophotometer (Hitachi, Tokyo, Japan).

3.4. Photocatalytic Tests

A 150 W xenon lamp with a cut off filter (≥420 nm) was used as the visible light source for photocatalysis. Using MO as model pollutants, photocatalyst (0.1 g) was suspended in MO solution (100 mL, 10 mg/L) with stirring. The solution was kept in the dark for 30 min to reach adsorption-desorption equilibrium between the organic molecules and the photocatalyst surface. Then, 4 mL reacted solutions in series were taken out and analyzed every 1 h. The concentrations of MO in the reacting solutions were analyzed by a Cary 500 UV-vis-NIR spectrophotometer (Varian, Palo Alto, CA, USA) at 464 nm.

4. Conclusions

In summary, using electrospinning technology and SILAR method, heterojunctions of p-BiOI/n-TiO$_2$ NFs have been successfully fabricated. Due to the p-n heterojunction effects and large specific surface area, the BiOI/TiO$_2$-C30 exhibits higher visible-light photocatalytic behavior in comparison with other samples for degradation of MO. Furthermore, the p-BiOI/n-TiO$_2$ NFs can be easily recycled without a decrease of the photocatalytic activity because of their nanofibrous nonwoven web structure property. It is expected that the p-BiOI/n-TiO$_2$ NFs with high photocatalytic activity will greatly promote their industrial application to eliminate the organic and inorganic pollutants from wastewater.

Supplementary Materials: Supplementary Materials: The following are available online at www.mdpi.com/1996-1944/9/2/90/s1.

Acknowledgments: Acknowledgments: The work is supported financially by the National Basic Research Program of China (973 Program) (Grant No. 2012CB933703), the National Natural Science Foundation of China (No. 91233204, 51272041, 61201107, and 51572045), the 111 Project (No. B13013), and the Fundamental Research Funds for the Central Universities (12SSXM001).

Author Contributions: Author Contributions: Kexin Wang and Xinghua Li performed the experiments and analyzed the data; Changlu Shao and Yichun Liu designed the experiments and revised the draft; Na Lu and Fujun Miao contribute reagents/materials/analysis tools; and Kexin Wang wrote the paper.

Conflicts of Interest: Conflicts of Interest: The authors declare no conflict of interest.

References

1. Hoffmann, M.R.; Martin, S.T.; Choi, W.; Bahnemann, D.W. Environmental applications of semiconductor photocatalysis. *Chem. Rev.* **1995**, *95*, 69–96.
2. Zhang, K.; Liu, C.; Huang, F.; Zheng, C.; Wang, W. Study of the electronic structure and photocatalytic activity of the BiOCl photocatalyst. *Appl. Catal. B Environ.* **2006**, *68*, 125–129.
3. Zhang, X.; Liu, X.; Fan, C.; Wang, Y.; Wang, Y.; Liang, Z. A novel BiOCl thin film prepared by electrochemical method and its application in photocatalysis. *Appl. Catal. B Environ.* **2013**, *132–133*, 332–341.
4. Zhang, X.; Ai, Z.H.; Jia, F.L.; Zhang, L.Z. Generalized one-pot synthesis, characterization, and photocatalytic activity of hierarchical BiOX (X = Cl, Br, I) nanoplate microspheres. *J. Phys. Chem. C* **2008**, *112*, 747–753.
5. Wang, D.H.; Gao, G.Q.; Zhang, Y.W.; Zhou, L.S.; Xu, A.W.; Chen, W. Nanosheet-constructed porous BiOCl with dominant {001} facets for superior photosensitized degradation. *Nanoscale* **2012**, *4*, 7780–7785.

6. Huo, Y.; Zhang, J.; Miao, M.; Jin, Y. Solvothermal synthesis of flower-like BiOBr microspheres with highly visible-light photocatalytic performances. *Appl. Catal. B Environ.* **2012**, *111–112*, 334–341.

7. Ye, L.Q.; Chen, J.N.; Tian, L.H.; Liu, J.Y.; Peng, T.Y.; Deng, K.J.; Zan, L. BiOI thin film via chemical vapor transport: Photocatalytic activity, durability, selectivity and mechanism. *Appl. Catal. B Environ.* **2013**, *130*, 1–7.

8. Cheng, H.; Wang, W.; Huang, B.; Wang, Z.; Zhan, J.; Qin, X.; Zhang, X.; Dai, Y. Tailoring AgI nanoparticles for the assembly of AgI/BiOI hierarchical hybrids with size-dependent photocatalytic activities. *J. Mater. Chem. A* **2013**, *1*, 7131–7136.

9. Xiao, X.; Zhang, W.-D. Facile synthesis of nanostructured BiOI microspheres with high visible light-induced photocatalytic activity. *J. Mater. Chem.* **2010**, *20*, 5866–5870.

10. Cao, S.; Chen, C.; Liu, T.; Tsang, Y.; Zhang, X.; Yu, W.; Chen, W. Synthesis of reduced graphene oxide/alpha-$Bi_2Mo_3O_{12}$@beta-Bi_2O_3 heterojunctions by organic electrolytes assisted UV-excited method. *Chem. Eng. J.* **2014**, *257*, 309–316.

11. Peng, Y.; Yan, M.; Chen, Q.G.; Fan, C.M.; Zhou, H.Y.; Xu, A.W. Novel onedimensional Bi_2O_3-Bi_2WO_6 p-n hierarchical heterojunction with enhanced photocatalytic activity. *J. Mater. Chem. A* **2014**, *2*, 8517–8524.

12. Chang, X.X.; Wang, T.; Zhang, P.; Zhang, J.J.; Li, A.; Gong, J.L. Enhanced surface reaction kinetics and charge separation of p-n heterojunction Co_3O_4/$BiVO_4$ photoanodes. *J. Am. Chem. Soc.* **2015**, *137*, 8356–8359.

13. Liu, Y.; Zhang, M.; Li, L.; Zhang, X. One-dimensional visible-light-driven bifunctional photocatalysts based on $Bi_4Ti_3O_{12}$ nanofiber frameworks and Bi2XO6 (X=Mo, W) nanosheets. *Appl. Catal. B Environ.* **2014**, *160–161*, 757–766.

14. Reddy, K.H.; Martha, S.; Parida, K.M. Fabrication of novel p-BiOI/n-$ZnTiO_3$ heterojunction for degradation of rhodamine 6G under visible light irradiation. *Inorg. Chem.* **2013**, *52*, 6390–6401.

15. Li, H.; Jin, Z.; Sun, H.; Sun, L.; Li, Q.; Zhao, X.; Jia, C.-J.; Fan, W. Facile fabrication of p-BiOI/n-Zn_2SnO_4 heterostructures with highly enhanced visible light photocatalytic performances. *Mater. Res. Bull.* **2014**, *55*, 196–204.

16. Jiang, J.; Zhang, X.; Sun, P.; Zhang, L. ZnO/BiOI heterostructures: Photoinduced charge-transfer property and enhanced visible-light photocatalytic activity. *J. Phys. Chem. C* **2011**, *115*, 20555–20564.

17. Hou, D.; Hu, X.; Hu, P.; Zhang, W.; Zhang, M.; Huang, Y. $Bi_4Ti_3O_{12}$ nanofibers-BiOI nanosheets p-n junction: Facile synthesis and enhanced visible-light photocatalytic activity. *Nanoscale* **2013**, *5*, 9764–9772.

18. Yoneyama, H.; Yamashita, Y.; Tamura, H. Heterogeneous photocatalytic reduction of dichromate on n-type semiconductor catalysts. *Nature* **1979**, *282*, 817–818.

19. Schneider, J.; Matsuoka, M.; Takeuchi, M.; Zhang, J.L.; Horiuchi, Y.; Anpo, M.; Bahnemann, D.W. Understanding TiO_2 photocatalysis: Mechanisms and materials. *Chem. Rev.* **2014**, *114*, 9919–9986.

20. Liu, M.; Li, H.; Zeng, Y.; Huang, T. Anatase TiO_2 single crystals with dominant {001} facets: Facile fabrication from Ti powders and enhanced photocatalytical activity. *Appl. Surf. Sci.* **2013**, *274*, 117–123.

21. Liu, M.; Piao, L.; Ju, S.; Lu, W.; Zhao, L.; Zhou, C.; Wang, W. Fabrication of micrometer-scale spherical titanate nanotube assemblies with high specific surface area. *Mater. Lett.* **2010**, *64*, 1204–1207.

22. Liu, M.; Zhong, M.; Li, H.; Piao, L.; Wang, W. Facile synthesis of hollow TiO_2 single nanocrystals with improved photocatalytic and photoelectrochemical activities. *ChemPlusChem* **2015**, *80*, 688–696.

23. Liu, M.; Piao, L.; Wang, W. Hierarchical TiO_2 spheres: Facile fabrication and enhanced photocatalysis. *Rare Metals* **2011**, *30*, 153–156.

24. Liu, M.; Sunada, K.; Hashimoto, K.; Miyauchi, M. Visible-light sensitive Cu(ii)-TiO_2 with sustained anti-viral activity for efficient indoor environmental remediation. *J. Mater. Chem. A* **2015**, *3*, 17312–17319.

25. Wu, D.Y.; Wang, H.Y.; Li, C.L.; Xia, J.; Song, X.J.; Huang, W.S. Photocatalytic self-cleaning properties of cotton fabrics functionalized with p-BiOI/n-TiO_2 heterojunction. *Surf. Coat. Technol.* **2014**, *258*, 672–676.

26. Li, Y.; Wang, J.; Liu, B.; Dang, L.; Yao, H.; Li, Z. BiOI-sensitized TiO_2 in phenol degradation: A novel efficient semiconductor sensitizer. *Chem. Phys. Lett.* **2011**, *508*, 102–106.

27. Zhang, D. Heterostructural BiOI/TiO_2 composite with highly enhanced visible light photocatalytic performance. *Russ. J. Phys. Chem. A* **2014**, *88*, 2476–2485.

28. Zhang, Z.; Shao, C.; Li, X.; Sun, Y.; Zhang, M.; Mu, J.; Zhang, P.; Guo, Z.; Liu, Y. Hierarchical assembly of ultrathin hexagonal SnS_2 nanosheets onto electrospun TiO_2 nanofibers: Enhanced photocatalytic activity based on photoinduced interfacial charge transfer. *Nanoscale* **2013**, *5*, 606–618.

29. Liu, C.; Wang, L.; Tang, Y.; Luo, S.; Liu, Y.; Zhang, S.; Zeng, Y.; Xu, Y. Vertical single or few-layer MoS_2 nanosheets rooting into TiO_2 nanofibers for highly efficient photocatalytic hydrogen evolution. *Appl. Catal. B Environ.* **2015**, *164*, 1–9.

30. Wang, K.X.; Shao, C.L.; Li, X.H.; Zhang, X.; Lu, N.; Miao, F.J.; Liu, Y.C. Hierarchical heterostructures of p-type BiOCl nanosheets on electrospun n-type TiO_2 nanofibers with enhanced photocatalytic activity. *Catal. Commun.* **2015**, *67*, 6–10.

31. Lu, M.X.; Shao, C.L.; Wang, K.X.; Lu, N.; Zhang, X.; Zhang, P.; Zhang, M.Y.; Li, X.H.; Liu, Y.C. p-MoO_3 nanostructures/n-TiO_2 nanofiber heterojunctions: Controlled fabrication and enhanced photocatalytic properties. *ACS Appl. Mater. Inter.* **2014**, *6*, 9004–9012.

32. Ye, L.; Tian, L.; Peng, T.; Zan, L. Synthesis of highly symmetrical BiOI single-crystal nanosheets and their {001} facet-dependent photoactivity. *J. Mater. Chem.* **2011**, *21*, 12479–12484.

33. Wang, S.M.; Guan, Y.; Wang, L.P.; Zhao, W.; He, H.; Xiao, J.; Yang, S.G.; Sun, C. Fabrication of a novel bifunctional material of BiOI/Ag_3VO_4 with high adsorption-photocatalysis for efficient treatment of dye wastewater. *Appl. Catal. B Environ.* **2015**, *168*, 448–457.

34. Liu, H.; Cao, W.; Su, Y.; Wang, Y.; Wang, X. Synthesis, characterization and photocatalytic performance of novel visible-light-induced Ag/BiOI. *Appl. Catal. B Environ.* **2012**, *111–112*, 271–279.

35. Chu, M.-W.; Ganne, M.; Caldes, M.T.; Brohan, L. X-ray photoelectron spectroscopy and high resolution electron microscopy studies of Aurivillius compounds: $Bi_{4-x}La_xTi_3O_{12}$(x = 0, 0.5, 0.75, 1.0, 1.5, and 2.0). *J. Appl. Phys.* **2002**, *91*, 3178.

36. Wei, X.X.; Chen, C.M.; Guo, S.Q.; Guo, F.; Li, X.M.; Wang, X.X.; Cui, H.T.; Zhao, L.F.; Li, W. Advanced visible-light-driven photocatalyst $BiOBr$-TiO_2-graphene composite with graphene as a nano-filler. *J. Mater. Chem. A* **2014**, *2*, 4667–4675.

37. Zhang, Y.C.; Yang, M.; Zhang, G.S.; Dionysiou, D.D. HNO_3-involved one-step low temperature solvothermal synthesis of N-doped TiO_2 nanocrystals for efficient photocatalytic reduction of Cr(VI) in water. *Appl. Catal. B Environ.* **2013**, *142*, 249–258.

Influence of Oxygen Concentration on the Performance of Ultra-Thin RF Magnetron Sputter Deposited Indium Tin Oxide Films as a Top Electrode for Photovoltaic Devices

Jephias Gwamuri, Murugesan Marikkannan, Jeyanthinath Mayandi,
Patrick K. Bowen and Joshua M. Pearce

Abstract: The opportunity for substantial efficiency enhancements of thin film hydrogenated amorphous silicon (a-Si:H) solar photovoltaic (PV) cells using plasmonic absorbers requires ultra-thin transparent conducting oxide top electrodes with low resistivity and high transmittances in the visible range of the electromagnetic spectrum. Fabricating ultra-thin indium tin oxide (ITO) films (sub-50 nm) using conventional methods has presented a number of challenges; however, a novel method involving chemical shaving of thicker (greater than 80 nm) RF sputter deposited high-quality ITO films has been demonstrated. This study investigates the effect of oxygen concentration on the etch rates of RF sputter deposited ITO films to provide a detailed understanding of the interaction of all critical experimental parameters to help create even thinner layers to allow for more finely tune plasmonic resonances. ITO films were deposited on silicon substrates with a 98-nm, thermally grown oxide using RF magnetron sputtering with oxygen concentrations of 0, 0.4 and 1.0 sccm and annealed at 300 °C air ambient. Then the films were etched using a combination of water and hydrochloric and nitric acids for 1, 3, 5 and 8 min at room temperature. In-between each etching process cycle, the films were characterized by X-ray diffraction, atomic force microscopy, Raman Spectroscopy, 4-point probe (electrical conductivity), and variable angle spectroscopic ellipsometry. All the films were polycrystalline in nature and highly oriented along the (222) reflection. Ultra-thin ITO films with record low resistivity values (as low as 5.83×10^{-4} Ω·cm) were obtained and high optical transparency is exhibited in the 300–1000 nm wavelength region for all the ITO films. The etch rate, preferred crystal lattice growth plane, d-spacing and lattice distortion were also observed to be highly dependent on the nature of growth environment for RF sputter deposited ITO films. The structural, electrical, and optical properties of the ITO films are discussed with respect to the oxygen ambient nature and etching time in detail to provide guidance for plasmonic enhanced a-Si:H solar PV cell fabrication.

Reprinted from *Materials*. Cite as: Gwamuri, J.; Marikkannan, M.; Mayandi, J.; Bowen, P.K.; Pearce, J.M. Influence of Oxygen Concentration on the Performance of Ultra-Thin RF Magnetron Sputter Deposited Indium Tin Oxide Films as a Top Electrode for Photovoltaic Devices. *Materials* **2016**, *9*, 63.

1. Introduction

Solar photovoltaic (PV) based electricity production is one of the significant ecofriendly methods to generate sustainable energy needed to mitigate the looming global energy crisis [1]. Despite technical improvements [2] and scaling [3], which have resulted in a significant reduction in crystalline silicon (c-Si) PV module costs, for continued PV industry growth [4,5], PV costs must continue to decline to reach a levelized cost of electricity [6] low enough to dominate the electricity market. One approach to reduced PV costs further is to transition to thin film PV technology [7]. Hydrogenated amorphous silicon (a-Si:H) based PV [8] have shown great potential for large scale [9] sustainable commercial production due to lower material costs and use of well-established fabrication techniques [10,11]. However, there is need to improve the efficiency of a-Si:H PV devices if they are to become the next dominant technology for solar cells commercialization. One method to improve a-Si:H PV performance is with optical enhancement [12]. Recent developments in plasmonic theory promise new light management methods for thin-film a-Si:H based solar cells [13–23]. However, previous work has shown these plasmonic approaches require the development of ultra-thin, low-loss and low-resistivity transparent conducting oxides (TCOs) [24]. Tin doped indium oxide (ITO), zinc oxide (ZnO) and tin oxide (SnO_2) are the three most important TCOs and are already widely used in the commercial thin film solar cells [25]. In addition, aluminum-dope zinc oxide (AZO) and fluorine-doped tin oxide (FTO) are among the other most dominant TCOs in various technological fields particularly the optoelectronic devices industry where TCOs have proved indispensable for applications such as photo electrochemical devices, light emitting diodes, liquid crystal displays and gas sensors [26,27]. ITOs can be prepared by direct current (DC) and radio frequency (RF) magnetron sputtering, electron beam evaporation, thermal vapor evaporation, spray pyrolysis, chemical solution deposition, and sol gel methods [28–34]. RF magnetron sputtering can be used to control the electrical and optical properties of the ITO thin films and is heavily used in industry [35].

Recent work by Vora *et al.* has emphasized the need for ultra-thin ITO top electrodes with low resistivity and high transmittances in the visible range of the electromagnetic spectrum as a prerequisite for the commercial realization of plasmonic-enhanced a-Si:H solar cells [36]. However, research by Gwamuri *et al.* has demonstrated that fabricating ultra-thin ITO films (sub-50 nm) using conversional methods presented a number of challenges since there is a trade-off between electrical and optical properties of the films [37]. It was evidenced from their results that electrical properties of RF sputter deposited sub-50 nm ITO films degraded drastically as their thickness is reduced, while the optical properties of the same films were seen to improve greatly [37]. To solve this problem, a novel method involving chemical shaving of thicker (greater than 80 nm) RF sputter deposited films was proposed and demonstrated [38]. Building on the promise of that technique, this study seeks to further understand the effect of oxygen concentration on the etch rates of RF sputter deposited ITO films and the impact on the TCO quality as a top electrode for PV devices. A detailed understanding of the interaction of all critical parameters, which determines the quality of ultra-thin ITO will help create even thinner layers with good quality to allow more finely tuned plasmonics resonances. ITO films were deposited using four different oxygen concentrations (0 sccm, 0.4 sccm, 1.0 sccm), annealed in air at 300 °C for 30 min and then etched for four different times (1, 3, 5 and 8 min) to establish the effect of oxygen on etch rates. These materials were characterized by X-ray diffraction (XRD), atomic force microscopy (AFM), Raman Spectroscopy, 4-point probe (4PP), and variable angle spectroscopic Ellipsometry (VASE). In addition, the thin films were investigated for candidates as acid-resistant TCOs for encapsulation of PV devices, which may reduce device processing steps and fabrication costs of completed modules in the future. The results are presented and discussed.

2. Materials and Methods

2.1. ITO Fabrication Process

ITO films were grown on (100) prime silicon substrates with a 98 nm thermally grown oxide, and on glass substrates using a 99.99% 100 mm diameter pressed ITO ($SnO_2:In_2O_3$ 10:90 wt%) target. Before the deposition the substrates were ultrasonically cleaned in isopropanol and in DI water for 15 min and dried using N_2 atmosphere. The sputtering chamber was initiated to a low 10^{-7} Torr base pressure and the pressure was maintained at 7.5×10^{-3} Torr. The distance between the target and substrates was kept constant at 75 mm. As a standard procedure, the target was pre-sputter cleaned at a power of 150 W, whereas the sputter deposition of the films was performed at 100 W. The argon gas flow rate was fixed at 10 sccm and the oxygen gas flow was varied such as 0, 0.4 and 1.0 sccm with sputter rate of 8–12 nm per minute. The sputter rate was seen to decrease with increase in oxygen flow rate. After deposition, ITO films were annealed at 300 °C for 30 min in air. ITO/Si films were subjected to the etching process using a standard chemical etchant mixture of HCl: HNO_3:H_2O (1:1:5) volume ratio. All the etching was performed at room temperature, resulting in a slow and controlled etch rate for the Si/SiO_2 films. Finally, the etched samples were thoroughly rinsed in DI water and dried under the nitrogen environment. This methodology was adapted from the previous study by Gwamuri et al., 2015 [37].

The ITO films processed under different argon-oxygen ambient were chemically etched and characterized using various tools. The structural analyses of the ITO films were carried out using X-ray diffraction (XRD-Scintag-2000 PTS, Scintag Inc., Cupertino, CA, USA). Raman spectra for the ultra-thin film samples were measured at room temperature using Jobin-Yvon LabRAM HR800 Raman Spectrometer (Horiba Scientific, Edison, NJ, USA) with the excitation wavelength of 633 nm and the resolution is about ~0.1 cm^{-1}. Sheet resistance of the samples was characterized using four point probe station consisting of ITO optimized tips with 500 micron tip radii set to 60 grams pressure and an RM3000 test unit from Jandel Engineering Limited, Kings Langley, UK. The optical transmission and thickness of the films was determined using variable angle spectroscopic ellipsometry (UV-VIS V-VASE with control module VB-400, J.A. Woollam Co., Lincoln, NE, USA). Surface roughness was evaluated using a Veeco Dimension 3000 atomic force microscope (Veeco, Oyster City, NY, USA) operated in tapping mode with Budget Sensors Tap300Al-G cantilevers (Innovative Solutions Bulgaria Ltd., Sofia, Bulgaria). It should be noted that transmittance data was measured for ITO on sodalime glass (SLG) substrate and all the rest of the data was on ITO on Si/SiO$_2$ substrate.

2.2. Chemical Shaving: Wet Etching

In this present work, the oxygen 0, 0.4 and 1.0 sccm deposited ITO films were used for the etching process for 1, 3, 5 and 8 min, respectively. The annealed ITO/Si samples are etched at room temperature using HCl:HNO$_3$:H$_2$O (1:1:5) combination and the resistivity and thickness of the films were checked for 1, 3, 5 and 8 min etched films. For the 0 sccm ITO films, the thickness of the film was changed from 70 to 44 nm for 1 to 5 min etching time. Similarly the 0.4 sccm films thickness changed from 89 to 47 nm and 84 to 22 nm for 1.0 sccm films. The decrement of thickness was reflected in the resistivity values. The chemical reaction of the HCl and HNO$_3$ etching reactions are as follows [39]:

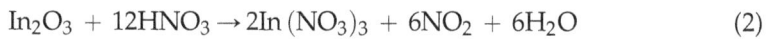

$$In_2O_3 + 2HCl \rightarrow 2InCl + H_2O + O_2 \, (\Delta H) \tag{1}$$

$$In_2O_3 + 12HNO_3 \rightarrow 2In \, (NO_3)_3 + 6NO_2 + 6H_2O \tag{2}$$

3. Results

3.1. Structural Analysis

3.1.1. XRD Analysis

XRD results for the ITO films deposited using different oxygen concentrations (0 sccm, 0.4 sccm, 1.0 sccm), annealed in air at 300 °C for 30 min and then etched for different times (1, 3, 5 and 8 min) are shown in Figure 1.

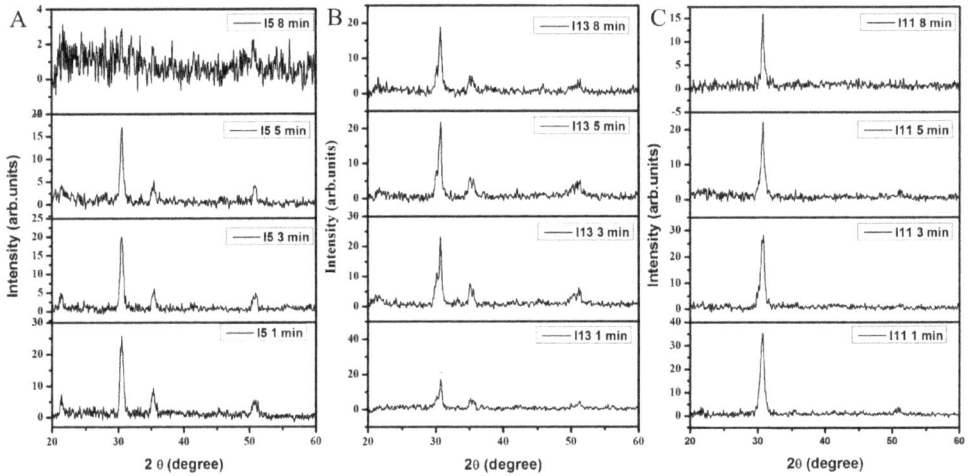

Figure 1. XRD pattern for ITO films deposited under different oxygen ambient conditions and etched for 1, 3, 5 and 8 min: (**A**) 0 sccm oxygen; (**B**) 0.4 sccm oxygen; (**C**) 1.0 sccm oxygen. Argon flow rate was maintained at 10 sccm for all materials.

In addition to that the peak shown at (222), (400) and (440) reflections are indexed to be cubic indium oxide (JCPDS No: 06-0416) [40]. All the films have a polycrystalline nature with stronger (222) reflection. No other tin phases could be identified from the cubic indium tin oxide. Normally the 30% of Sn is needed to exhibiting the SnO_2 diffraction lines in ITO. The (222) and (400) plane is ascribed for oxygen efficient and deficient nature of ITO films [41]. The effect of the oxygen flow rate on the peak intensity of the ITO films is clearly shown in the XRD spectrum. There is a general increase in the peak intensities with increased oxygen flow rate. Similarly the reflections such as (211), (400) and (440) are due to the minimum oxygen concentration in the sputter chamber. These planes are absent in the XRD pattern of ITO film processed in an oxygen-rich (1.0 sccm) environment. There is a strong evidence that for the highest oxygen ambient (1.0 sccm), (222) is the preferred growth orientation for RF sputter deposited ITO films. Varying the oxygen concentration will result in changing the preferred growth orientation of the films to other crystal lattice planes such as the (211), (400) or (440). The intensity ratios are strongly dependent on the critical level of In^{3+} and O^{2-} pairs and the pairs' density is different for different etching periods of time [42]. The presence of high oxygen concentration induce the In-O bonding networks formation and promote growth of the (222) crystal lattice planes.

During the etching, ITO films are reduced to In–Cl and In-$(NO_3)_3$ resulting in the change in crystallinity of films etched for different periods of time. The structural parameters such as d spacing, lattice constants, net lattice distortion and grain sizes are estimated and listed in Table 1 in comparison to data from the Joint Committee on Powder Diffraction Standards (JCPDS)/International Centre for Diffraction Data (ICDD) database.

The etching process also distorts the ITO structural long-range order, which has an impact on the opto-electronic properties of the films. The grain size of films did not change even after etching for 8 min., particularly for ITO films processed in an oxygen deficient environment. During the etching process the excess weakly bound oxygen atoms are removed from the ITO surfaces exposing layers with different grain sizes. The ITO structure distortion due to etching for longer periods of time (8 min) can be seen from the XRD spectra shown in Figure 1. There was however no evidence of ITO film for the results shown in Figure 1A after they were etched for 8 min. There is evidence of decreased crystallinity for the rest of the ITO films (Figure 1B,C) as the oxygen atoms are stripped from the In–O network by the HCl and HNO_3.

Table 1. Structural parameters of ITO sputtered films with 0, 0.4 and 1.0 sccm oxygen and etched at 1, 3, 5 and 8 min.

Oxygen Flow Rate (sccm)	Etching Time (min)	D Spacing (222) (Å)	Lattice Constant (222) (Å)	Net-Lattice Distortion	Grain Size (222) (nm)
Standard JCPDS for ITO 06-0416	–	2.921	10.1180	–	–
0	1	2.932	10.1552	–0.0036	16
	3	2.934	10.1629	–0.2970	16
	5	2.934	10.1629	–0.2885	17
	8	–	–	–	–
0.4	1	2.908	10.0731	–0.0075	31
	3	2.912	10.0869	–0.0157	25
	5	2.914	10.0954	–0.0153	23
	8	2.914	10.0954	–0.0169	20
1.0	1	2.917	10.1059	–	13
	3	2.913	10.0915	–0.2828	13
	5	2.911	10.0845	–	14
	8	2.908	10.0764	–	19

3.1.2. AFM Analysis

Figure 2 shows the AFM surface topology of the ITO films deposited under three different oxygen environments and etched for 1 min and 8 min.

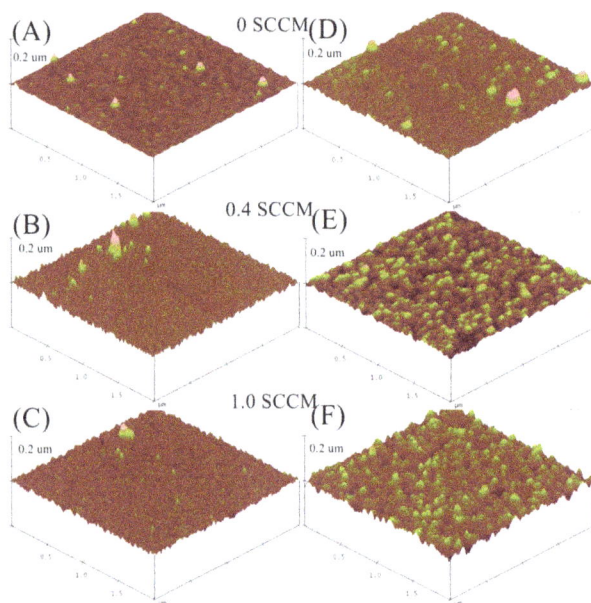

Figure 2. Surface topology image for 2 μm × 2 μm × 0.2 μm of the ITO film deposited under various oxygen environments: (**A**) 0 sccm oxygen; (**B**) 0.4 sccm oxygen; (**C**) 1.0 sccm oxygen, and etched for 8 min; (**D**) 0 sccm oxygen; (**E**) 0.4 sccm oxygen and (**F**) 1 sccm oxygen. (**A**–**C**) etched for 1 min and (**D**–**F**) films etched for 8 min. The etching was performed at room temperature.

Figure 2A,B shows the ITO films deposited in an oxygen deficient ambient and etched for 1 and 8 min, respectively. Spherical sized grains are clearly visible in all AFM images presented in Figure 2. There is variation of surface roughness of the films with both oxygen flow rate and etching time of the ITO films. The minimum value of surface roughness of 0.65 nm was measured for ITO films sputtered using 0.4 sccm oxygen flow rate and etched for 1 min, while a maximum surface roughness value of 8.9 nm was observed for films processed at 1.0 sccm oxygen flow rate and etched for 1 min. There was a slight increase in roughness with etching time observed for 0 sccm and 0.4 sccm ITO film, for etching times 1 min to 8 min. However, the 1.0 sccm films, showed the greatest variation in surface roughness even after 1 min etching process. Generally, the surface roughness of the films are observed to increase when the oxygen gas concentration is increased during processing.

3.1.3. Raman Spectroscopy

Figure 3 shows the Raman spectrum for ITO deposited at various oxygen compositions and etched at 1, 3, 5 and 8 min respectively. Raman spectroscopy is used to determine the structural conformations of the materials. Group theory predicts the Raman modes for cubic indium oxide, such as 4Ag (Raman), 4Eg (Raman), 14Tg (Raman), 5Au (inactive), and 16Tu (infrared) modes [43]. The modes observed are at 303, 621 and 675 cm^{-1} for all the films. Noticeable modes are exhibited at 302 and 621 for Eg and In–O vibrational mode [44]. The observed Raman modes in Figure 3 are in good agreement with previous reported results [40]. There are no other additional modes observable for the SnO and SnO$_2$ structures. In addition to that the broad band shown at 976 to 1013 cm^{-1} for all the etched films and it was not unassignable. The peak appeared at 1132, 1112, 1097 and 1120 cm^{-1} for 0, 0.4 and 1.0 sccm ITO etched films. These peaks are reported in the commercially ITO films [45]. The Raman results are correlated with XRD results. No other mixed phases were observed in the Raman spectrum indicating that etching process had no or little effect on the ITO structure.

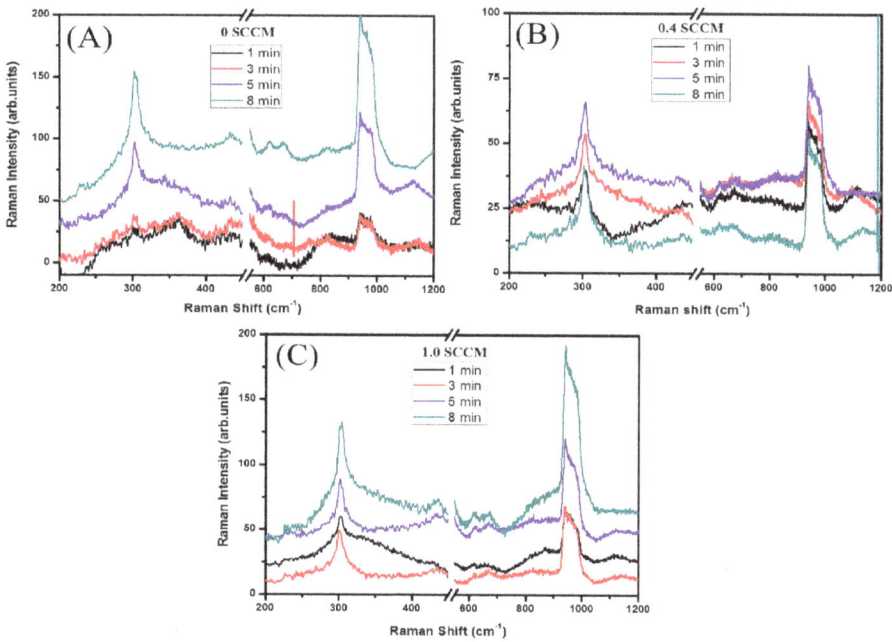

Figure 3. Raman spectra for the ITO films deposited under various oxygen concentrations and etched for 1, 3, 5 and 8 min., respectively. (**A**) 0 sccm; (**B**) 0.4 sccm; (**C**) 1.0 sccm.

3.2. Resistivity

The electrical properties of the different oxygen ambient deposited and etched ITO films were measured using a four point probe. The sheet resistance values of the ITO films are changed with respect to the oxygen ambient nature and etching time and are summarized in Table 2.

Table 2. Electrical and optical parameters of ITO films deposited under various oxygen compositions and etched for 1, 3, 5 and 8 min.

Oxygen Flow Rate (sccm)	Etching Time (min)	Sheet Resistance (Ω/square)	Thickness (nm)	Resistivity ($\Omega \cdot$cm)	Transmission (%)
	1	83.28	70	5.83×10^{-4}	76.29
0	3	103.47	59	6.11×10^{-4}	93.98
	5	209.49	44	9.22×10^{-4}	90.27
	8	–	–	–	100
	1	209.02	89	1.86×10^{-3}	91.45
0.4	3	194.23	88	1.71×10^{-3}	85.26
	5	240.08	85	2.04×10^{-3}	84.85
	8	326.9	47	1.54×10^{-3}	83.71
	1	1000	84	8.4×10^{-3}	90.96
1.0	3	2000	62	1.24×10^{-2}	89.25
	5	2400	50	1.20×10^{-2}	100
	8	7350	22	1.62×10^{-2}	100

From the obtained results, the minimum sheet resistance was observed for ITO deposited using argon ambient (0 sccm oxygen) and etched for 1 min. However, the same films exhibited the worst transmittance of about 76%. During processing in an argon rich environment, the bombardment by argon neutrals creates dangling bonds in the substrates and created the oxygen vacancies in the ITO films [46]. The argon environment (10 sccm) (*i.e.,* the oxygen deficient environment) promotes oxygen vacancies that enhance electrical resistivity while degrading the optical properties of the films. This is reflected in the XRD spectra, where the (400) and (440) lattice planes are enhanced for ITO films processed in low oxygen (0 sccm and 0.4 sccm) environments. Hence, the 1.0 sccm deposited films in which the (222) lattice plane is dominant, showed a higher electrical resistivity compared to the other films. The resistivity of the films are highly dependent on the film thickness, which is a function of the etching time. Increasing the etching time decreases the thickness of the films, and, hence, the electrical properties while improving optical properties.

3.3. Transmittance

Optical transmittances of the ITO films on glass substrates are recorded from 300 to 1000 nm at room temperature and shown in Figure 2. All the films exhibited the highest average optical transmittance in the higher wavelength range. The highest optical transmittance is attained for 0sccm oxygen ITO film etched for 8 min with and average etch rate of 5.2 nm/min (for 5 min etch) and the 1.0 sccm films etched for 5 and 8 min with average etch rates of 5.25 and 7.75 nm/min, respectively. The results are summarized in Table 2. The thickness of the film is an important parameter for determining both electrical and optical properties of the ITO films. In this work, the thickness was quantified using spectroscopic ellipsometry measurements and is shown in the Table 2. For transmittance measurements, the ITO films were deposited on SLG substrates. The SLG transmittance is measured and used as baseline data. All ITO films transmittance data involve baseline subtraction, hence 100% transmittance means that all of the ITO film has been etched off. The etch rates were much faster for the ITO on glass such that all the film was etched-off after an 8-min etch (Figure 4A), and 5 and 8 min (Figure 4C).

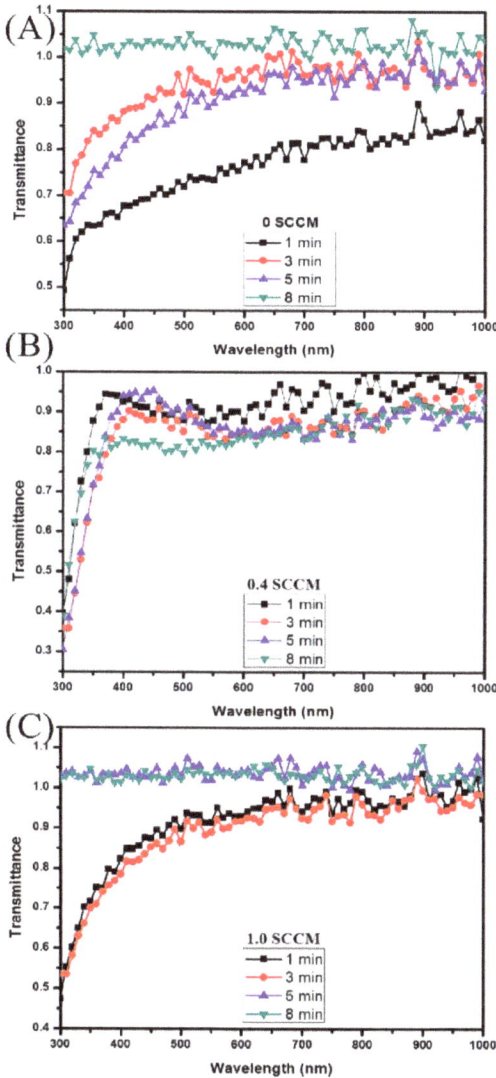

Figure 4. Optical transmission spectrum for the RF sputter deposited ITO films at different oxygen compositions and etched for min (**A**) 0 sccm; (**B**) 0.4 sccm; and (**C**) 1.0 sccm.

The results show a direct correlation between the oxygen concentration and the optical transmittances of the films and an inverse relationship with the electrical conductivity of the ITO films. These results are in agreement with observation from previous studies [47,48]. Figure 4A for 8 min etch, and Figure 4C for 5 and 8 min etch showed a transmittance cut-off wavelength around 350 nm indicating the absorption edge.

4. Discussion

The results presented provide further insight on the interaction of the most common fabrication variables that influence the electrical and optical properties of ITO films for PV and other opto-electronic applications. There is evidence of a strong correlation between oxygen concentration and both the resistivity and transmittances of RF sputter deposited ITO films. The processing conditions have a strong bearing on the structure of the ITO films. The (222) lattice planes are preferred in films grown in oxygen rich ambient whilst more lattice planes; (211), (222), (400), and (440) are observed in films grown under less oxygen or oxygen deficient conditions. Wan *et al.* have reported the (211), (400) and (440) planes reflections as being associated with ITO films processed under high RF power in an oxygen deficient atmosphere [35]. From the structural analysis, the exhibited (222) reflection clearly indicated the cubic indium oxide formation. The film growth rate decreased with increased O_2 concentration resulting in a much thicker ITO critical thickness (amorphous to polycrystalline transition thickness) for the 0 sccm RF sputtered films. The overall film is mixed phase crystalline and amorphous in nature. The noise is due to the ultra-thin porous and amorphous film left once the top crystalline film is etched off. This is not observed in the 0.4 and 1 sccm films because the increased oxygen composition results in reduced growth rates giving films that are more crystalline in nature with a much thinner critical thickness.

The lattice parameters and lattice distortion are seen to vary closely with oxygen concentration in the sputter chamber and the length of the etching process. The different oxygen–argon ratios sputtered ITO films have different etching behaviors, which then effected their electrical and optical properties. During the etching process, the crystal lattice of ITO films is degraded due to the exchange of bonds between indium oxide with HCl and HNO_3. This means that the indium oxide In–O and H–Cl bonds are substituted by In–Cl, In-$(NO_3)_3$ and O–H in the ITO surfaces [49]. These kinds of reactions may reduce the oxygen concentrations and distort structural long range order of the ITO films. This was reflected in the variation of electrical and optical properties of the ITO films with etching time. As no evidence of tin phases were detected, it can be concluded the reactions involving tin phases have negligible effect on the overall etch rates described in this study.

It is interesting to note that the ITO films processed at the 0.4 sccm oxygen flow rate presented the greatest resistance to acid etching in addition to exhibiting above moderate electrical and optical properties. These films show potential as candidate materials for encapsulation of PV devices or transparent conducting electrodes for varied application in acid-rich environment. However, further research into optimization of anti-acid (acid resistant) ITO films will be required before the material can be implemented in commercial PV devices.

Usually ITO is sputtered in varied combinations of reactive gas environments of argon with oxygen, hydrogen and nitrogen [40]. The oxygen ambient has been shown to be an important parameter to control electrical and optical properties. The highest oxygen concentrations enhance the transmission property and the oxygen deficient nature (oxygen vacancies) increased the electrical conductivity of the ITO thin films [47,48]. Hence, a sufficient amount of oxygen concentration can improve the opto-electronic performance of ITO thin films. High-quality ultra-thin ITO films are a needed significant step towards the realization and possible commercialization of plasmonic-based a-Si:H thin-film PV devices [15,24,36–38]. These devices have a potential to transform the thin-film based solar cells industry due to their low cost and ease of fabrication. In addition, plasmonic-enhanced PV has the potential to exhibit sophisticated light management schemes enabling unprecedented control over the trapping and propagation of light within the active region of the PV device [15], which would be expected to result in record-high device solar energy conversion efficiencies.

5. Conclusions

In this study, ultra-ITO thin films have been RF sputter deposited using different oxygen flow rates and chemical shaving is performed at room temperature for different time periods. The thicknesses of the films are altered as a result from 89 nm to 22 nm. In-between each etching process cycle, the films were characterized for both electrical and optical properties. Generally, the transmittance of the ITO films was observed to increase with decreasing film thickness, while the electrical properties were observed to degrade for the same films. This was attributed to the distortion of the In–O lattice long-range order due to the reduction reaction between the ITO and the etchants (acids). The novel method of chemical shaving further investigated here, is a simple and low-cost method with the potential to produce low loss and highly conductive ultra-thin and acid resistant ITO films for applications ranging from PV devices transparent electrodes to anti-acid materials. Using this method, ultra-thin ITO films with record low resistivity values (as low as 5.83×10^{-4} $\Omega{\cdot}cm$) were obtained and the optical transmission is generally high in the 300–1000 nm wavelength region for all films. The etching rate strongly depends on the oxygen concentrations of RF sputtered ITO films as well as on the post process annealing.

This processing has an effect on the oxygen vacancies densities even for the 0 sccm O_2 films. Surface roughness increased as the concentration of oxygen increased as expected. The etching reactions are simple redox reaction, hence the rates should increase with increases in O_2 concentration especially for non-stoichiometric films with distorted ITO matrix. The etch rate, preferred crystal lattice growth plane, d-spacing and lattice distortion were also observed to be highly dependent on the nature of growth environment for RF sputter deposited ITO films.

Acknowledgments: Acknowledgments: This work was supported by the National Science Foundation under grant award number CBET-1235750 and the Fulbright S&T award. Jeyanthinath Mayandi thanks the University Grants Commission (UGC) for providing support through Raman fellowship 2014–2015 to visit Michigan Technological University, Houghton, MI 49931, USA.

Author Contributions: Author Contributions: Jephias Gwamuri, Jeyanthinath Mayandi and Joshua M. Pearce conceived and designed the experiments; Jephias Gwamuri, Jeyanthinath Mayandi, Murugesan Marikkannan and Patrick K. Bowen performed the experiments; all authors analyzed the data; Jeyanthinath Mayandi and Joshua M. Pearce contributed materials and tools; and all authors wrote the paper.

Conflicts of Interest: Conflicts of Interest: The authors declare no conflict of interest.

Abbreviations

4-PP	Four Point Probe
AFM	Atomic Force Microscopy
a-Si:H	Hydrogenated Amorphous Silicon
AZO	Aluminum doped Zinc Oxide
c-Si	Crystalline Silicon
DC	Direct Current
FTO	Fluorine doped-Tin oxide
ITO	Indium Tin Oxide
PV	Photovoltaic
RF	Radio Frequency
SCCM	Standard Cubic Centimeters per Minute (flow unit)
SLG	Sodaline Glass
SnO_2	Tin Oxide
TCOs	Transparent Conducting Oxides
VASE	Variable Angle Spectroscopic Ellipsometry
XRD	X-ray Diffraction
ZnO	Zinc Oxide

References

1. Pearce, J.M. Photovoltaics—A path to sustainable futures. *Futures* **2002**, *34*, 663–674.
2. Pillai, U. Drivers of cost reduction in solar photovoltaic. *Energy Econ.* **2015**, *50*, 286–293.
3. Honeyman, C.; Kimbis, T. *Solar Market. Insight Report 2014 Q2*; Solar Energy Industrial Association and GTM research: Washington, DC, USA, 2014.
4. Candelise, C.; Winskel, M.; Gross, R.J.K. The dynamics of solar PV costs and prices as a challenge for technology forecasting. *Renew. Sustain. Energy Rev.* **2013**, *26*, 96–107.
5. Rubin, E.S.; Azevedo, I.M.L.; Jaramillo, P.; Yeh, S. A review of learning rates for electricity supply technologies. *Energy Policy* **2015**, *86*, 198–218.
6. Branker, K.; Pathak, M.J.M.; Pearce, J.M. A Review of Solar Photovoltaic Levelized Cost of Electricity. *Renew. Sustain. Energy Rev.* **2011**, *15*, 4470–4482.
7. Shah, A.; Torres, P.; Tscharner, R.; Wyrsch, N.; Keppner, H. Photovoltaic technology: The case for thin-film solar cells. *Science* **1999**, *285*, 692–698.
8. Carlson, D.E.; Wronski, C.R. Amorphous Silicon Solar Cell. *Appl. Phys. Lett.* **1976**, *28*, 671–673.
9. Pearce, J.M. Industrial symbiosis of very large-scale photovoltaic manufacturing. *Renew. Energy* **2008**, *33*, 1101–1108.
10. Wronski, C.R.; Pearce, J.M.; Koval, R.J.; Ferlauto, A.S.; Collins, R.W. Progress in Amorphous Silicon Based Solar Cell Technology. Available online: http://www.rio12.com/rio02/proceedings/pdf/067_Wronski.pdf (accessed on 13 October 2015).
11. Collins, R.W.; Ferlauto, A.S.; Ferreira, G.M.; Chen, C.; Koh, J.; Koval, R.J.; Lee, Y.; Pearce, J.M.; Wronski, C.R. Evolution of microstructure and phase in amorphous, protocrystalline, and microcrystalline silicon studied by real time spectroscopic ellipsometry. *Solar Energy Mater. Solar Cells* **2003**, *78*, 143–180.
12. Deckman, H.W.; Wronski, C.R.; Witzke, H.; Yablonovitch, E. Optically enhanced amorphous silicon solar cells. *Appl. Phys. Lett.* **1983**, *42*, 968–970.
13. Atwater, H.A.; Polman, A. Plasmonics for improved photovoltaic devices. *Nat. Mater.* **2010**, *9*, 205–213.
14. Derkacs, D.; Lim, S.H.; Matheu, P.; Mar, W.; Yu, E.T. Improved performance of amorphous silicon solar cells via scattering from surface plasmon polaritons in nearby metallic nanoparticles. *Appl. Phys. Lett.* **2006**, *89*, 093103–093105.
15. Gwamuri, J.; Güney, D.Ö.; Pearce, J.M. Advances in Plasmonic Light Trapping in Thin-Film Solar Photovoltaic Devices. In *Solar Cell. Nanotechnology*; Tiwari, A., Boukherroub, R., Maheshwar Sharon, M., Eds.; Wiley: Hoboken, NJ, USA, 2013; pp. 241–269.
16. Spinelli, P.; Ferry, V.E.; van de Groep, J.; van Lare, M.; Verschuuren, M.A.; Schropp, R.E.I.; Atwater, H.A.; Polman, A. Plasmonic light trapping in thin-film Si solar cells. *J. Opt.* **2012**, *14*, 024002–024012.
17. Cai, W.; Salaev, V.M. *Optical Metamaterials: Fundamentals and Applications*, 1st ed.; Springer: New York, NY, USA, 2010; p. 278.

18. Maier, S.A.; Atwater, H.A. Plasmonics: Localization and guiding of electromagnetic energy in metal/dielectric structures. *J. Appl. Phys.* **2005**, *98*, 011101–011110.

19. Aydin, K.; Ferry, V.E.; Briggs, R.M.; Atwater, H.A. Broadband polarization-independent resonant light absorption using ultrathin plasmonic super absorbers. *Nat. Commun.* **2011**, *2*, 517.

20. Wu, C.; Avitzour, Y.; Shvets, G. Ultra-thin wide-angle perfect absorber for infrared frequencies. *Proc. SPIE* **2008**, *7029*.

21. Ferry, V.E.; Verschuuren, M.A.; van Lare, C.; Ruud, E.I.; Atwater, H.A.; Polman, A. Optimized spatial correlations for broadband light trapping nano patterns in high efficiency ultrathin film a-Si:H solar cells. *Nano Lett.* **2011**, *11*, 4239–4245.

22. Trevino, J.; Forestiere, C.; Di Martino, G.; Yerci, S.; Priolo, F.; Dal Negro, L. Plasmonic-photonic arrays with aperiodic spiral order for ultra-thin film solar cells. *Opt. Express* **2012**, *20*, A418–A430.

23. Massiot, I.; Colin, C.; Pere-Laperne, N.; Roca i Cabarrocas, P.; Sauvan, C.; Lalanne, P.; Pelouard, J.-L.; Collin, S. Nanopatterned front contact for broadband absorption in ultra-thin amorphous silicon solar cells. *Appl. Phys. Lett.* **2012**, *101*, 163901–163903.

24. Vora, A.; Gwamuri, J.; Pala, N.; Kulkarni, A.; Pearce, J.M.; Güney, D.Ö. Exchanging Ohmic losses in metamaterial absorbers with useful optical absorption for photovoltaics. *Sci. Rep.* **2014**, *4*, 1–13.

25. Sato, K.; Gotoh, Y.; Wakayama, Y.; Hayashi, Y.; Adachi, K.; Nishimura, H. Highly textured SnO_2:F TCO films for a-Si solar cells. *Rep. Res. Lab. Asahi Glass Co. Ltd.* **1992**, *42*, 129–137.

26. Dixit, A.; Sudakar, C.; Naik, R.; Naik, V.M.; Lawes, G. Undoped vacuum annealed In_2O_3 thin films as a transparent conducting oxide. *Appl. Phys. Lett.* **2009**, *95*, 192105–192107.

27. Lan, J.H.; Kanicki, J. ITO surface ball formation induced by atomic hydrogen in PECVD and HW-CVD tools. *Thin Solid Films* **1997**, *304*, 123–129.

28. Thøgersen, A.; Rein, M.; Monakhov, E.; Mayandi, J.; Diplas, S. Elemental distribution and oxygen deficiency of magnetron sputtered indium tin oxide films. *J. Appl. Phys.* **2011**, *109*, 113532.

29. Park, H.K.; Yoon, S.W.; Chung, W.W.; Min, B.K.; Do, Y.R. Fabrication and characterization of large-scale Multifunctional transparent ITO nanorod films. *J. Mater. Chem. A* **2013**, *1*, 5860–5867.

30. Castaneda, S.I.; Rueda, F.; Diaz, R.; Ripalda, J.M.; Montero, I. Whiskers in Indium tin oxide films obtained by electron beam evaporation. *J. Appl. Phys.* **1998**, *83*, 1–8.

31. Yao, J.L.; Hao, S.; Wilkinson, J.S. Indium Tin Oxide Films by Sequential Evaporation. *Thin Solid Films* **1990**, *189*, 221–233.

32. Kobayashi, H.; Kogetsu, Y.; Ishida, T.; Nakato, Y. Increase in photovoltage of "indium tin oxide/Silicon oxide/ mat-textured n–silicon" junction solar cells by silicon peroxidation and annealing processes. *J. Appl. Phys.* **1993**, *74*, 4756–4761.

33. Lee, J.; Lee, S.; Li, G.; Petruska, M.A.; Paine, D.C.; Sun, S. A Facile Solution-Phase Approach to Transparent and Conducting ITO Nanocrystal Assemblies. *J. Am. Chem. Soc.* **2012**, *134*, 13410–13414.

160

34. Chen, Z.; Li, W.; Li, R.; Zhang, Y.; Xu, G.; Cheng, H. Fabrication of Highly Transparent and Conductive Indium-Tin Oxide Thin films with a high figure of merit via solution processing. *Langmuir* **2013**, *29*, 13836–13842.

35. Wan, D.; Chen, P.; Liang, J.; Li, S.; Huang, F. (211)-Orientation Preference of Transparent Conducting In$_2$O$_3$:Sn Films and Its Formation Mechanism. *ACS. Appl. Mater. Interfaces* **2011**, *3*, 4751–4755.

36. Vora, A.; Gwamuri, J.; Pearce, J.M.; Bergstrom, P.L.; Guney, D.O. Multi-resonant silver nano-disk patterned thin film hydrogenated amorphous silicon solar cells for Staebler-Wronski effect compensation. *J. Appl. Phys.* **2014**, *116*, 093103.

37. Gwamuri, J.; Vora, A.; Khanal, R.R.; Phillips, A.B.; Heben, M.J.; Guney, D.O.; Bergstrom, P.; Kulkarni, A.; Pearce, J.M. Limitations of ultra-thin transparent conducting oxides for integration into plasmonic-enhanced thin-film solar photovoltaic devices. *Mater. Renew. Sustain. Energy* **2015**, *4*, 1–12.

38. Gwamuri, J.; Vora, A.; Mayandi, J.; Guney, D.O.; Bergstrom, P.; Pearce, J.M. A New Method of Preparing Highly Conductive Ultra-Thin Indium Tin Oxide for Plasmonic-Enhanced Thin Film Solar Photovoltaic Devices. **2016**, to be published.

39. Hang, C.J.; Su, Y.K.; Wu, S.L. The effect of solvent on the etching of ITO electrode. *Mater. Chem. Phys.* **2004**, *84*, 146–150.

40. Marikkannan, M.; Subramanian, M.; Mayandi, J.; Tanemura, M.; Vishnukanthan, V.; Pearce, J.M. Effect of ambient combinations of argon, oxygen, and hydrogen on the properties of DC magnetron sputtered indium tin oxide films. *AIP Adv.* **2015**, *5*, 017128–017138.

41. Luo, S.; Kohiki, S.; Okada, K.; Shoji, F.; Shishido, T. Hydrogen effects on crystallinity, photoluminescence, and magnetization of indium tin oxide thin films sputter-deposited on glass substrate without heat treatment. *Phys. Status Solidi A* **2010**, *207*, 386–390.

42. Kato, K.; Omoto, H.; Tomioka, T.; Takamatsu, A. Changes in electrical and structural properties of indium oxide thin films through post-deposition annealing. *Thin Solid Films* **2011**, *520*, 110–116.

43. Liu, D.; Lei, W.W.; Zou, B. High pressure X-ray diffraction and Raman spectra study of indium oxide. *J. Appl. Phys.* **2008**, *104*, 083506–083511.

44. Berengue, O.M.; Rodrigues, A.D.; Dalmaschio, C.J.; Lanfredi, A.J.C.; Leite, E.R.; Chiquito, A.J. Structural characterization of indium oxide nanostructures: A Raman analysis. *J. Phys. D Appl. Phys.* **2010**, *43*, 045401–045404.

45. Chandrasekhar, R.; Choy, K.L. Innovative and cost-effective synthesis of indium tin oxide films. *Thin Solid Films* **2001**, *398–399*, 59–64.

46. Luo, S.N.; Kono, A.; Nouchi, N.; Shoji, F. Effective creation of oxygen vacancies as an electron carrier source in tin-doped indium oxide films by plasma sputtering. *J. Appl. Phys.* **2006**, *100*, 113701–113709.

47. Okada, K.; Kohiki, S.; Luo, S.; Sekiba, D.; Ishii, S.; Mitome, M.; Kohno, A.; Tajiri, T.; Shoji, F. Correlation between resistivity and oxygen vacancy of hydrogen-doped indium tin oxide thin films. *Thin Solid Films* **2011**, *519*, 3557–3561.

48. Ashida, T.; Miyamuru, A.; Oka, N.; Sato, Y.; Yagi, T.; Taketoshi, N.; Baba, T.; Shigesato, Y. Thermal transport properties of polycrystalline tin-doped indium oxide films. *J. Appl. Phys.* **2009**, *105*, 073709–073712.

49. Van den Meerakker, J.E.A.M.; Baarslag, P.C.; Walrave, W.; Vink, T.J.; Daams, J.L.C. On the homogeneity of sputter-deposited ITO films Part II. Etching behavior. *Thin Solid Films* **1995**, *266*, 152–156.

Efficiency Enhancement of Dye-Sensitized Solar Cells' Performance with ZnO Nanorods Grown by Low-Temperature Hydrothermal Reaction

Fang-I Lai, Jui-Fu Yang and Shou-Yi Kuo

Abstract: In this study, aligned zinc oxide (ZnO) nanorods (NRs) with various lengths (1.5–5 µm) were deposited on ZnO:Al (AZO)-coated glass substrates by using a solution phase deposition method; these NRs were prepared for application as working electrodes to increase the photovoltaic conversion efficiency of solar cells. The results were observed in detail by using X-ray diffraction, field-emission scanning electron microscopy, UV-visible spectrophotometry, electrochemical impedance spectroscopy, incident photo-to-current conversion efficiency, and solar simulation. The results indicated that when the lengths of the ZnO NRs increased, the adsorption of D-719 dyes through the ZnO NRs increased along with enhancing the short-circuit photocurrent and open-circuit voltage of the cell. An optimal power conversion efficiency of 0.64% was obtained in a dye-sensitized solar cell (DSSC) containing the ZnO NR with a length of 5 µm. The objective of this study was to facilitate the development of a ZnO-based DSSC.

Reprinted from *Materials*. Cite as: Lai, F.-I.; Yang, J.-F.; Kuo, S.-Y. Efficiency Enhancement of Dye-Sensitized Solar Cells' Performance with ZnO Nanorods Grown by Low-Temperature Hydrothermal Reaction. *Materials* **2015**, *8*, 8860–8867.

1. Introduction

Dye-sensitized solar cells (DSSC) belong to the third generation of solar cells. Due their low-cost materials and low-cost technologies, they are the promising replacement for conventional silicon-based solar cells [1]. The highest single-cell conversion efficiency of 13% is comparable to the Si cells [2]. Generally, TiO_2 nanoparticle films coated onto fluorine-doped tin oxide (FTO) layers are made as the photoelectrode in DSSCs because of their suitable chemical affinity and surface area for dye adsorption as well as their proper energy band promising charge transfer between the electrolytes and dye [3,4]. However, the one problem of DSSCs is that not all of the photogenerated electrons can arrive at the collecting electrode, because electron transport within the nanoparticle network takes place via a series of hops to adjacent particles, and the energy damage that occurs during charge transport processes results in conversion efficiency. This trapping process results in the transport becoming slow, and an increase in scattering, which greatly increases the

163

recombination of the electrons with the oxidized dye molecules, reducing efficiency and oxidized redox species. In order to enhance dye adsorption, the thickness of TiO_2 should be increased. However, this recombination problem is aggravated in TiO_2 nanocrystals by reason of a depletion layer on the TiO_2 nanocrystallite surface, and its severity increases as the photoelectrode film thickness increases [5]. In response to this problem, the paper proposes a ZnO-based DSSC technology as a replacement for TiO_2 in solar cells. Zinc oxide has received a great deal of attention as a photoanode in dye-sensitized solar cells (DSSCs) due to its large exciton-binding energy (60 meV) and large band gap (3.37 eV) [6]. Furthermore, its electron mobility is higher than that of TiO_2 by two-to-three orders of magnitude [7]. Therefore, ZnO is anticipated to demonstrate faster electron transport as well as decreased recombination damage compared to TiO_2. Nevertheless, studies have reported that the entire efficiency of TiO_2 DSSCs is higher than that of ZnO DSSCs. The efficiency of TiO_2 thin-passivation shell layers is higher than the highest reported efficiency of ZnO DSSCs [8], in which the principal problem is the dye adsorption process in ZnO DSSCs. Because of the high carboxylic acid binding groups in the dyes, the dissolution of ZnO and precipitation of dye-Zn^{2+} complexes occurs. This phenomenon results in a poor overall electron injection efficiency of the dye [9].

Several approaches exist for enhancing the efficiency of ZnO DSSCs. One method is to introduce a surface passivation layer to a mesoporous ZnO framework; nevertheless, this may aggravate the dye adsorption problems. Alternatively, conventional particulate structures can be changed by replacing the internal surface area and morphology of the photoanode. Nevertheless, the surface area and diffusion length are incompatible. Augmenting the photoanode thickness empowered a higher number of dye molecules to be fixed; this, however, increases the possibility of electron recombination because of the extended distance through which electrons diffuse to the transparent conductive oxide (TCO) collector. This trapping process results in augmented scattering and slows down the electron transport which increases the recombination of the electrons with the oxidized redox species or the oxidized dye molecules, hence reducing efficiency. One probable strategy for ameliorating electron transport in DSSCs is to supersede the nanoparticle photoelectrode with a single-crystalline nanorod (or nanosheet, nanobelt, nanotip) photoelectrode. Electrons can be led through a direct electron path within a nanorod rather than by multiple-scattering transport between nanoparticles. In research, the electron transport is tens to hundreds of times slower in nanoparticle DSSCs than in nanorod-based DSSCs [10–12]. Therefore, many works have been performed on the synthesis of TiO, and ZnO nanostructures for applications in DSSCs [13–15].

However, the utilization of FTO may not be the best method for improving the cell performance. One problem is that the small difference in the work function between ZnO and FTO does not supply sufficient driving force for the charge injection from the ZnO nanowires to FTO, which hints that new TCO materials should be used in ZnO-based DSSCs. Lee *et al.* use the ZnO:Al (AZO) film to replace the FTO layer as the TCO layer [4]. Their structure was accomplished by a three-step process, TCO, seed layer, and nanostructure, but this method was slight complicated. To simplify the procedures, we used a two-step process in this study, and present a detailed discussion. These characteristics were observed using X-ray diffraction (XRD), UV-visible spectrophotometry, field-emission scanning electron microscopy (FE-SEM), electrochemical impedance spectroscopy (EIS), incident photon-to-electron conversion efficiency (IPCE), and solar simulation.

2. Experimental

Figure 1 illustrates the schematic structures of DSSCs with ZnO nanorods of various lengths, which are shown in Figure 1. First, radio-frequency sputtering was used to deposit a ZnO:Al (AZO) seed layer (approximately 300 nm) on Corning-glass substrates with a sheet resistance of 18 Ω/sq, and the defined area of the seed layer was 1 cm^2. The Pt (H$_2$PtC$_{l6}$ solid content: <6%, viscosity: ~50 cps, eversolar Pt-100) film was also deposited on the AZO/Corning-glass substrates by spin-coating. These substrates were used for growing ZnO NRs. The ZnO NRs were deposited using zinc nitrate (Zn(NO$_3$)$_2$6H$_2$O, Aldrich) and hexamethylenetetrasece (C$_6$H$_{12}$N$_4$, HMT, Aldrich). Both mixtures were melted in deionized water to a concentration of 0.02 M and stored at 90 °C for 9 h. These solutions were replaced every 9 h, and the corresponding ZnO NRs were denoted by 18- and 27-h NRs. The hydrothermal chemical reactions for the ZnO NRs are expressed as follows:

$$C_6H_{12}N_4 + 6H_2O \rightarrow 6HCHO + 4NH_3 \tag{1}$$

$$NH_3 + H_2O \rightarrow NH_4^+ + OH^- \tag{2}$$

$$Zn^{2+} + 2OH^- \rightarrow Zn(OH)_2 \tag{3}$$

$$Zn(OH)_2 \underset{heat}{\rightarrow} ZnO + H_2O \tag{4}$$

After the reaction was complete, the resulting ZnO NRs were rinsed with deionized water to remove residual ZnO particles and impurities. A D-719 dye, *cis*-bis(isothiocyanato)bis(2,2'-bipyridyl-4,4'-dicarboxylato) ruthenium(II)bis-tetrabutylammonium, (Everlight Chemical Industrial Corp., Taipei, Taiwan) was dissolved in acetonitrile for preparing a 0.5 mM dye solution. Dye sensitization was propagated by soaking the ZnO photoelectrodes in the D-719 dye at room temperature for 2 h. A sandwich-type configuration was used to measure the presentation of the DSSCs. An active area of 1 cm^2 was assembled by using a Pt-coated AZO substrate as a counter electrode, and the Pt/AZO was heated at 200 °C for 30 min in air. The DSSC was sealed employing a polymer resin (Surlyn) to act as a spacer. The electrolyte was injected into the space among the electrodes from these two holes, and then these two holes were sealed completely by using Surlyn. The electrolyte (0.5 M 4-tert-butyl-pyridine + 0.05 M I$_2$ + 0.5 M LiI + 0.6 M tetrabutylammonium iodide) was injected to the cell and then sealed with UV gel. The influence of growth time on the structural and optical properties of these ZnO NRs was analyzed by XRD and UV-visible spectrophotometry. Surface morphologies of the ZnO nanorods were examined using field-emission scanning electron microscope (FESEM). The photocurrent-voltage (I-V) characteristic curves were measured using Keithley 2420 under AM 1.5 illumination. The electrochemical impedance spectroscopy (EIS) was measured under the light illumination of AM 1.5 G (100 mW/cm^2) with an impedance analyzer (Autolab PGSTAT 30) (Metrohm Autolab, Utrecht, Netherlands) when a device was applied with its open-circuit voltage (Voc). An additional alternative sinusoidal voltage amplitude 10 mV was also applied between an anode and a cathode of a device over the frequency range of 0.02~100 kHz. The external quantum efficiency (EQE) results were acquired from a system using a 300 W xenon lamp (Newport 66984) light source and a monochromator (Newport 74112) (Newport Corporation, Taipei, Taiwan). The beam spot size at the sample measured was approximately 1 mm × 3 mm. The temperature was controlled at 25 °C during the measurements.

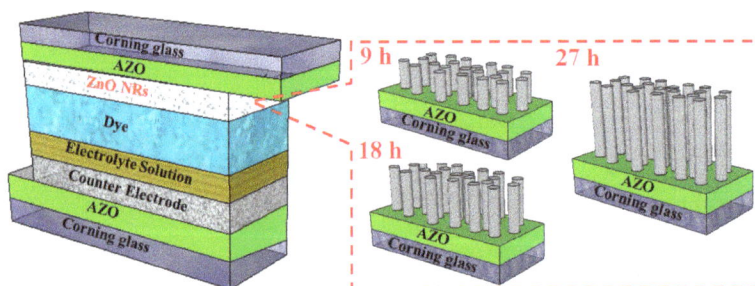

Figure 1. The schematic illustrations of DSSCs with ZnO nanorods.

3. Results and Discussion

In this study, ZnO NRs with various lengths were grown on AZO substrates of photoanodes to increase the optical absorption of the dye. Figure 2a shows the respective XRD patterns for the ZnO NRs derived from the 9-, 18-, and 27-h reactions, respectively. The crystalline structure was analyzed using XRD measurements according to a θ/2θ configuration. In principle, the XRD spectra indicate that the ZnO films developed without the presence of secondary phases or groups. All the samples have a hexagonal wurtzite structure of ZnO and grew along the c-axis; this enabled the observation of the ZnO (002) diffraction plane in the XRD pattern. The increase in intensity of the diffraction peak and also the narrowing of the peak, in other words, decrease in the full width at half maximum (FWHM) of the peak, with the length of ZnO NRs increased, and the crystallinity improvement of the ZnO NRs. Existing dye uptake measurements were based on dye desorption from the photoanode after a specified 30 min using a NaOH solution, and the succeeding UV-Vis spectroscopy. For the quantitative analysis of dye loading, the washing course for desorption of dye from the anodes was performed using the known volume of 0.1 mM NaOH aqueous solution. The dye detached from the ZnO NRs as implemented for different lengths of ZnO NRs in literature. Figure 2b illustrates the absorptions of solutions containing 0.01 mM dye, indicating that dyes detached from the ZnO NRs at 9- (black line), 18- (red line), and 27-h (blue line), respectively. The area of both films was 1 cm^2. The results depicted in Figure 2b can be used to calculate the dye loadings and light absorptions at 530 nm (the peak dye absorption) for the ZnO NRs from the 9-, 18-, and 27-h reactions. The lengths of the 18- and 27-h ZnO NRs were longer than those of the 9-h ZnO NRs, and they demonstrated an improvement in light harvesting and dye loading with increased NRs.

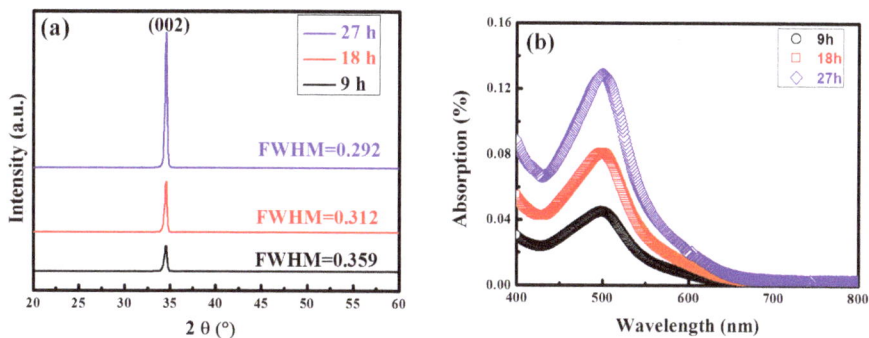

Figure 2. (a) XRD patterns of ZnO nanorods grown with different duration; (b) Optical absorption spectra of D-719 dye detached from the ZnO NRs with various lengths.

As mentioned, ZnO NRs with various lengths were grown on AZO substrates, and these NRs were used in DSSCs (Figure 3). Figure 3a–f illustrate FE-SEM images of the respective ZnO NRs from the 9-, 18-, and 27-h reactions grown on the AZO substrates, indicating that the ZnO NRs were adequately grown on substrates with a distinctive, clear morphology. Furthermore, the diameters, lengths, and

167

aspect ratios of the NRs were in the range of 76–110 nm, 1.5–5 μm, and 20.7–47.9, respectively. Greene *et al.* indicated that the growing temperature influences the upright growth of ZnO NRs [16].

Figure 4a depicts the Nyquist plots of the impedance spectra. To characterize the AZO/dye/electrolyte interface, the open-circuit voltage (Voc) levels of the DSSCs were evaluated under AM 1.5 illumination by conducting EIS measurements. The Nyquist plots indicate a small semicircle at high frequencies and a large semicircle at low frequencies. The inset in Figure 4a shows the equivalent circuit. Usually, all the spectra of the DSSCs exhibit three semicircles, which are ascribed to the electrochemical reaction at the Pt counter electrode, charge transfer at the TiO_2/dye/electrolyte, and Warburg diffusion process of I^-/I^{3-}, respectively [17,18]. In the present study, the charge transfer resistance at the ZnO/dye/electrolyte interface (Rct_2) decreased when the aspect ratio of the ZnO NRs was varied from 20.7 to 47.6. This may be attributable to the increase in the diameter size, length, and quality of ZnO NRs, which led to an increase in the dye adsorption as well as penetration of electron mobility into the pores of the AZO electrode (Figure 4a). The better collected and transported electrons had a lower possibility of recombination, and the electron lifetime was increased [19]. Figure 4b shows Bode phase plots indicating the characteristic frequency peaks ($1–10^4$ Hz). The characteristic frequency peak shifted to a lower frequency when the aspect ratio increased, and the characteristic frequency can be considered as the inverse of the electron lifetime (τ_e) or recombination lifetime (τ_r) in an AZO film [20,21]. This implies that the NRs with an aspect ratio of 47.6 (grown for 27 h) had the longest electron lifetime in the AZO film. The results indicate that the ZnO NRs, which were grown for 27 h (aspect ratio: 47.6), on the AZO film had a lower transport resistance and a longer electron lifetime in the AZO electrode. The electron lifetimes in the AZO films increased from 3.25 to 6.12 ms when the aspect ratio increased from 20.7 to 47.6. This result is consistent with the following results obtained from cell performance and EIS analysis.

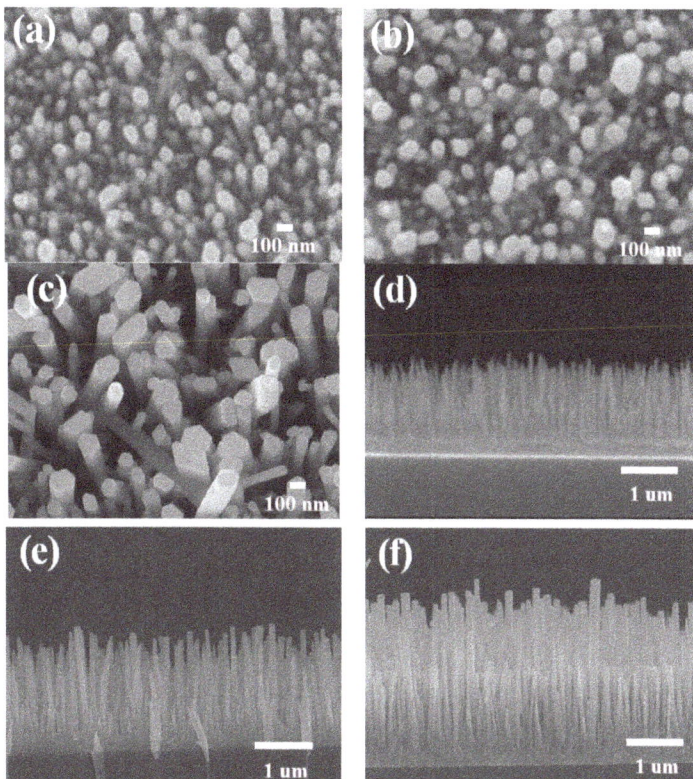

Figure 3. SEM images of ZnO nanorods fabricated under various growth time. (**a–c**) Top-view images of ZnO nanorods grown at 9 h, 18 h, and 27 h; (**d–f**) Side-view images of ZnO nanorods grown at 9 h, 18 h, and 27 h, respectively.

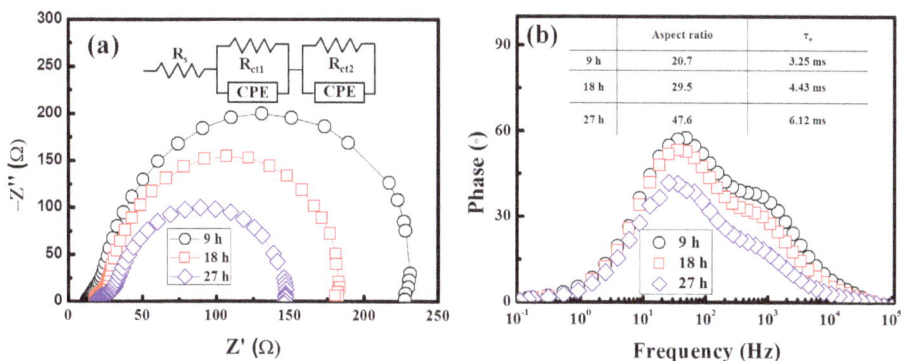

Figure 4. Electrochemical impedance spectra of DSSCs containing ZnO nanorods with various lengths. (**a**) Nyquist plots and (**b**) Bode phase plots. The equivalent circuit of this study is shown in the inset of (**a**).

169

Figure 5a shows the J-V curve for the DSSCs containing the ZnO NRs obtained from the 9-, 18-, and 27-h reactions, indicating that the short-circuit current density (Jsc) and cell performance significantly increase with the NR length. As revealed in the figures, the photovoltaic performances of our DSSCs employing ZnO NRs are comparable to published literature [22–24]. A higher amount of dye was adsorbed on longer NRs than on shorter NRs, indicating that longer NRs improve photon absorption and carrier generation. These results indicate that cell performance is strongly dependent on the electrode surface area. Increasing the NR length results in a larger surface area, which leads to a higher adsorption of dyes as well as a higher conversion efficiency. Furthermore, the Voc of the longer ZnO NRs was higher than that of the shorter ZnO NRs. This higher Voc is attributable to a reduction in recombination losses at ZnO/dye interfaces. Regarding the performance of the cells containing the ZnO NRs grown for various periods, the cell containing the 27-h ZnO NRs demonstrated optimal performance with a conversion efficiency (η) of 0.64%, Voc of 0.62 V, Jsc of 2.56 mA/cm^2, and fill factor of 0.42. The NRs also provide direct pathways from the point of photogeneration to the conducting substrate. These pathways ensure the rapid collection of carriers generated throughout the device. Figure 5b depicts the IPCE spectra of the DSSCs (D-719 dye) containing the 9-, 18-, and 27-h ZnO NRs, indicating a strong peak at 520 nm; this peak is attributable to the characteristic excitations of the D-719 dye. Our ZnO-based DSSCs show poor conversion efficiencies when compared to conventional TiO$_2$-based DSSCs, as shown in the inset of Figure 5a. The main reason is the corrosion of ZnO on reacting with an acid and the low amounts of dyes that are adsorbed during the production. During the process, an amount of Zn^{2+} ions are dissolved into the solution from the surface of the ZnO nanorods. Subsequently, aggregation of Zn^{2+} ions with sensitizer dyes occurs, and the phenomenon was reported for several organic sensitizer dyes as well as ruthenium complexes [25,26]. Once aggregation takes place in DSSCs, the power conversion efficiency will dramatically decrease [27]. Despite the lower efficiencies in our ZnO-based DSSCs, the use of ZnO nanorods still shows high potential because of its better crystallinity and higher electron mobility. To overcome the chemical instability of ZnO, the introduction of non-ruthenium-based sensitizers and the utilization of different nanotechnological architectures of ZnO might be practical approaches.

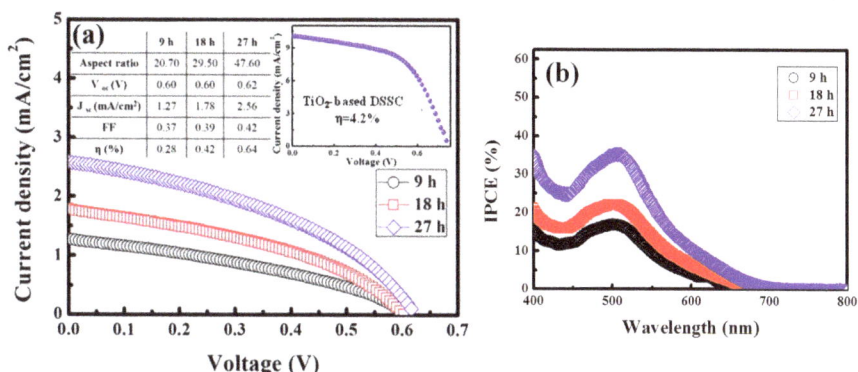

Figure 5. (**a**) J-V measurements under AM 1.5 illumination (100 mA·cm^{-2}) and (**b**) IPCE spectra of DSSCs containing ZnO nanorods grown at various durations. Shown in the inset of Figure 5a is the photovoltaic performance of DSSC employing TiO$_2$ nanoparticles.

4. Conclusions

In this study, we prepared ZnO NRs, with a two-step process which is simple and easy, for use as photoanodes in DSSCs. Moreover, the results reveal that DSSCs containing longer ZnO NRs demonstrate higher photovoltaic performance than DSSCs containing shorter ZnO NRs. Compared with shorter ZnO NRs, longer ZnO NRs exhibit a larger surface area, which enables efficient dye loading and light harvesting, reduced charge recombination, and faster electron transport. These improvements enhanced power conversion for application in DSSCs.

Acknowledgments: Acknowledgments: The authors would like to thank the experimental support from Hsiu-Po Kuo in Chang Gung University. This work was supported by the Green Technology Research Center of Chang Gung University and the Ministry of Science and Technology of Taiwan under contract numbers MOST104-2622-E-182-003-CC3 and MOST104-2112-M-182-005.

Author Contributions: Author Contributions: Fang-I Lai and Jui-Fu Yang designed and carried out the experiment and statistical analysis and participated in drafting the manuscript. Shou-Yi Kuo supervised the research and revised the manuscript. All authors read and approved the final manuscript.

Conflicts of Interest: Conflicts of Interest: The authors declare no conflict of interest.

References

1. O'Regan, B.; Grätzel, M. A low-cost, high-efficiency solar cell based on dye-sensitized colloidal TiO$_2$ films. *Nature* **1991**, *353*, 737–740.

2. Yella, A.; Lee, H.W.; Tsao, H.N.; Yi, C.; Chandiran, A.K.; Nazeeruddin, Md.K.; Diau, E.W.G.; Yeh, C.Y.; Zakeeruddin, S.M.; Grätzel, M. Porphyrin-sensitized solar cells with cobalt (II/III)-based redox electrolyte exceed 12 percent efficiency. *Science* **2011**, *334*, 629–634.

3. Wang, N.; Han, L.; He, H.; Park, N.H.; Koumoto, K. A novel high-performance photovoltaic–thermoelectric hybrid device. *Energy Environ. Sci.* **2011**, *4*, 3676–3679.

4. Lee, S.H. Al-doped ZnO Thin Film: A new transparent conducting layer for ZnO nanowire-based dye-sensitized solar cells. *J. Phys. Chem. C* **2010**, *114*, 7185–1789.

5. Lee, C.J.; Lee, T.J.; Lyu, S.C.; Zhang, Y.; Ruh, H.; Lee, H.J. Field emission from well-aligned zinc oxide nanowires grown at low temperature. *Appl. Phys. Lett.* **2002**, *81*, 3648–3650.

6. Yagi, E.; Hasiguti, R.R.; Aono, M. Electronic conduction above 4 K of slightly reduced oxygen-deficient rutile TiO_{2-x}. *Phys. Rev. B Condes. Matter* **1996**, *54*, 7945–7956.

7. Zhang, Q.; Chou, T.P.; Russo, B.; Jenekhe, S.A.; Cao, G. Aggregation of ZnO nanocrystallites for high conversion efficiency in dye-sensitized solar cell. *Angew. Chem. Int. Ed. Engl.* **2008**, *47*, 2402–2406.

8. Quintana, M.; Edvinsson, T.; Hagfeldt, A.; Boschloo, G. Comparison of dye-sensitized ZnO and TiO_2 solar cells: Studies of charge transport and carrier lifetime. *J. Phys. Chem. C* **2007**, *111*, 1035–1041.

9. Wong, D.K.P.; Ku, C.H.; Chen, Y.R.; Chen, G.R.; Wu, J.J. Enhancing electron collection efficiency and effective diffusion length in dye-sensitized solar cells. *Chemphyschem* **2009**, *10*, 2698–2702.

10. Baxter, J.B.; Aydil, E.S. Dye-sensitized solar cells based on semiconductor morphologies with ZnO nanowires. *Sol. Energy Mater. Sol. Cells* **2006**, *90*, 607–622.

11. Zhu, K.; Neale, N.R.; Miedaner, A.; Frank, A.J. Enhanced charge-collection efficiencies and light scattering in dye-sensitized solar cells using oriented TiO_2 nanotubes arrays. *Nano Lett.* **2007**, *7*, 69–74.

12. Yodyingyong, S.; Zhang, Q.F.; Park, K.; Dandeneau, C.S.; Zhou, X.Y.; Triampo, D.; Cao, G.Z. ZnO nanoparticles and nanowire array hybrid photoanodes for dye-sensitized solar cells. *Appl. Phys. Lett.* **2010**, *96*, 073115.

13. Jeng, M.J.; Wung, Y.L.; Chang, L.B.; Chow, L. Dye-sensitized solar cells with anatase TiO_2 nanorods prepared by hydrothermal method. *Int. J. Photoenergy* **2013**, *2013*, 280253.

14. Choopun, S.; Tubtimtae, A.; Santhaveesuk, T.; Nilphai, S.; Wongrat, E.; Hongsith, N. Zinc oxide nanostructures for applications as ethanol sensors and dye-sensitized solar cells. *Appl. Surf. Sci.* **2009**, *256*, 998–1002.

15. Xie, Y.L.; Li, Z.X.; Xu, Z.G.; Zhang, H.L. Preparation of coaxial TiO_2/ZnO nanotube arrays for high-efficiency photo-energy conversion applications. *Electrochem. Commun.* **2011**, *13*, 788–791.

16. Greene, L.; Law, M.; Tan, D.H.; Goldberger, J.; Yang, P. General route to vertical ZnO nanowire arrays using textured ZnO seeds. *Nano Lett.* **2005**, *5*, 1231–1236.

17. Longo, C.; Freitas, J.; de Paoli, M.A. Performance and stability of TiO$_2$/dye solar cells assembled with flexible electrodes and a polymer electrolyte. *J. Photochem. Photobiol. A Chem.* **2003**, *159*, 33–39.

18. Bernard, M.C.; Cachet, H.; Falaras, P.; Hugot-Le Goff, A.; Kalbac, M.; Lukes, I.; Oanh, N.T.; Stergiopoulos, T.; Arabatzis, I. Sensitization of TiO$_2$ by polypyridine dyes: Role of the electron donor. *J. Electrochem. Soc.* **2003**, *150*, E155–E164.

19. Lin, L.Y.; Yeh, M.H.; Lee, C.P.; Chou, C.Y.; Ho, K.C. Flexible dye-sensitized solar cells with one-dimensional ZnO nanorods as electron collection centers in photoanodes. *Electrochimica Acta* **2013**, *88*, 421–428.

20. Schlichthorl, G.; Park, N.G.; Frank, A.J. Evaluation of the charge-collection efficiency of dye-sensitized nanocrystalline TiO$_2$ solar cells. *J. Phys. Chem. B* **1999**, *103*, 782–791.

21. Kern, R.; Sastrawan, R.; Ferber, J.; Stangl, R.; Luther, J. Modeling and interpretation of electrical impedance spectra of dye solar cells operated under open-circuit conditions. *Electrochimica Acta* **2002**, *47*, 4213–4225.

22. Chen, H.; Du Pasquier, A.; Saraf, G.; Zhong, J.; Lu, Y. Dye-sensitized solar cells using ZnO nanotips and Ga-doped ZnO films. *Semicond. Sci. Technol.* **2008**, *23*, 045004.

23. Afifi, A.; Tabatabaei, M.K. Efficiency investigation of dye-sensitized solar cells based on the zinc oxide nanowires. *Orient. J. Chem.* **2014**, *30*, 155–160.

24. Hsu, Y.F.; Xi, Y.Y.; Djurišić, A.B.; Chen, W.K. ZnO nanorods for solar cells: Hydrothermal growth *versus* vapor deposition. *App. Phys. Lett.* **2008**, *92*, 133507.

25. Baumeler, R.; Rys, P.; Zollinger, H.; Hel, V. Über die spektrale Sensibilisierung von Zinkoxid: II. Anomales Sorptionsverhalten von o, o′-Dihydroxyazofarbstoffen und Entladungskinetik des damit sensibilisierten Zinkoxids. *Chim. Acta* **1973**, *56*, 2450–2460.

26. Hauffe, K.H. On spectral sensitization of zinc oxide. *Photogr. Sci. Eng.* **1976**, *20*, 124–134.

27. Keis, K.; Lindgren, J.; Lindquist, S.E.; Hagfeldt, A. Studies of the adsorption process of Ru complexes in nanoporous ZnO electrodes. *Langmuir.* **2000**, *16*, 4688–4694.

Enhanced Erbium-Doped Ceria Nanostructure Coating to Improve Solar Cell Performance

Nader Shehata, Michael Clavel, Kathleen Meehan, Effat Samir, Soha Gaballah and Mohammed Salah

Abstract: This paper discusses the effect of adding reduced erbium-doped ceria nanoparticles (REDC NPs) as a coating on silicon solar cells. Reduced ceria nanoparticles doped with erbium have the advantages of both improving conductivity and optical conversion of solar cells. Oxygen vacancies in ceria nanoparticles reduce Ce^{4+} to Ce^{3+} which follow the rule of improving conductivity of solar cells through the hopping mechanism. The existence of Ce^{3+} helps in the down-conversion from 430 nm excitation to 530 nm emission. The erbium dopant forms energy levels inside the low-phonon ceria host to up-convert the 780 nm excitations into green and red emissions. When coating reduced erbium-doped ceria nanoparticles on the back side of a solar cell, a promising improvement in the solar cell efficiency has been observed from 15% to 16.5% due to the mutual impact of improved electric conductivity and multi-optical conversions. Finally, the impact of the added coater on the electric field distribution inside the solar cell has been studied.

Reprinted from *Materials*. Cite as: Shehata, N.; Clavel, M.; Meehan, K.; Samir, E.; Gaballah, S.; Salah, M. Enhanced Erbium-Doped Ceria Nanostructure Coating to Improve Solar Cell Performance. *Materials* **2015**, *8*, 7663–7672.

1. Introduction

Optical nanostructures that emit visible light when excited by ultra-violet (UV) or infrared (IR) photons have been extensively studied for solar energy applications [1,2]. Recent research on one of these nanomaterials, cerium oxide (ceria) nanoparticles, has shown that its material properties are extremely well suited for a lot of applications [3–6]. Visible emission from either UV excitation (down-conversion) or IR excitation (up-conversion) can be obtained from ceria nanoparticles. However, both up- and down-conversion processes involve different physiochemical properties in ceria and optimization of each optical process via various nanoparticle synthesis and post-growth procedures tends to quench the efficiency of the other process.

Coating solar cells or panels with nanostructures has been recently investigated to enhance the conversion efficiency of the cells [7,8]. In this paper, it is aimed to coat a polycrystalline silicon cell with a thin layer of reduced erbium-doped ceria nanoparticles to improve the cell efficiency. The synthesized reduced erbium-doped ceria nanoparticles (REDC NPs) would have two main characteristics: to have higher conductivity and to be applicable for optical up- and down-conversions. In detail, the synthesized doped ceria nanoparticles would have relatively high concentrations of tri-valent cerium ions in trap states with a higher concentration of oxygen vacancies. When coated on the solar cells, the synthesized reduced ceria could have higher conductivity and improve the mobility of the generated photoelectrons, due to the increased rate of cerium ion conversion from +4 to +3 states accompanied with an increasing creation rate of charged O-vacancies. In addition, the reduced erbium-doped ceria nanoparticles have the unique material properties to act as an optical medium for both down-conversion and up-conversion at the same time to generate multi-wavelength visible emissions under near-UV and IR excitations, respectively. "Reduced" means that the nanoparticles are synthesized under a reduction environment using hydrogen. This environment helps to form oxygen vacancies and cerium ions (+3 states). These cerium tri-valent trap states are responsible for optical down-conversion. However, "non-reduced" means that there is no hydrogen during synthesis, which would not form the Ce^{3+} states. Then, without the reduction environment, the erbium-doped ceria nanoparticles are abbreviated EDC NPs. Then, the used synthesis process results in a high concentration of Ce^{3+} ions associated with the oxygen vacancies in ceria, which is required to obtain high fluorescence efficiency in the down-conversion process. Simultaneously, the synthesized nanoparticles contain the molecular energy levels of erbium that are required for up-conversion. Therefore, REDC NPs which are synthesized using this procedure can emit visible light when excited with either or both UV or IR photons. The synthesized nanoparticles were analyzed using optical absorbance spectroscopy, direct band gap calculations, fluorescence spectroscopy, transmission electron microscope (TEM), X-ray diffraction (XRD) and the electrical conductivity measurement. Then, the synthesized reduced nanoparticles were coated on polycrystalline silicon cells for improving the cell efficiency, which has been proved through I–V analysis, in addition to other cell characteristics such as open circuit voltage, short circuit current, and fill factor. Also, rate generation and E-field distributions of the coated cell were analyzed. Compared to other ceria nanostructure coatings in the literature [9–11], our novel coating offers the simultaneous enhancement of both optical and conductive properties which leads to improving the solar cell's efficiency without considering the traditional anti-reflection coatings. In addition, our synthesized nanoparticles are relatively low-cost and easy to prepare with a simple chemical synthesis procedure with a simple coating technique.

2. Results and Discussion

2.1. Nanoparticles Characterization

The optical absorption spectra of the synthesized REDC NPs are plotted in Figure 1a. Consequently, the corresponding values for the calculated allowed direct band gaps of the annealed samples are shown in Figure 1b, using Equation (1) [12,13].

$$\propto E = A\left(E - E_g\right)^{1/2} \tag{1}$$

where α is the absorbance coefficient, A is a constant dependent on the effective masses of electrons and holes of the material, E is the absorbed photon energy, and E_g is the direct allowed band gap. Thus, from the absorbance dispersion results in Figure 1a, $(\propto E)^2$ is presented *versus* the photon energy (E) as shown in Figure 1b. Then, the intersection between the extension of the linear region of the resulted curve with the E-axis indicates the band gap. Compared to the non-reduced nanoparticles, it can be observed that the band gap of the annealed nanoparticles is biased towards 3 eV, which is approximately the band gap energy for Ce_2O_3. Thus, there is evidence for the formation of a higher concentration of Ce^{3+} with corresponding oxygen vacancies [14]. The annealed REDC NPs are imaged using TEM as shown in Figure 2a. The mean diameter is found to be 10 nm, which shows that our synthesized REDC NPs are smaller than other optical nanoparticles that have been studied as an optical active medium for down- or up-conversion [15,16], which can lead to both better conductive and optical properties due to a higher surface-to-volume ratio. The XRD pattern is presented in Figure 2b, measured on a sample of the REDC NPs annealed at 700 °C, to demonstrate that the predominant nanostructure of the REDC NPs is cerium dioxide [17,18]. From the measurement of the width of individual intensity peaks, the average size (t_{XRD}) of nanoparticles, or the diameter in case of spherical-shaped nanoparticles, can be calculated from Scherrer's equation [13]

$$t_{XRD} = \frac{0.9\lambda}{\beta cos\theta} \tag{2}$$

where λ is the wavelength of the incident X-rays (0.15406 nm), β is the full-width half-maximum (FWHM), and θ is the diffraction angle. From the first peak, which represents the most stable plane of ceria (111), the average size of the nanoparticles is found to be ~10 nm.

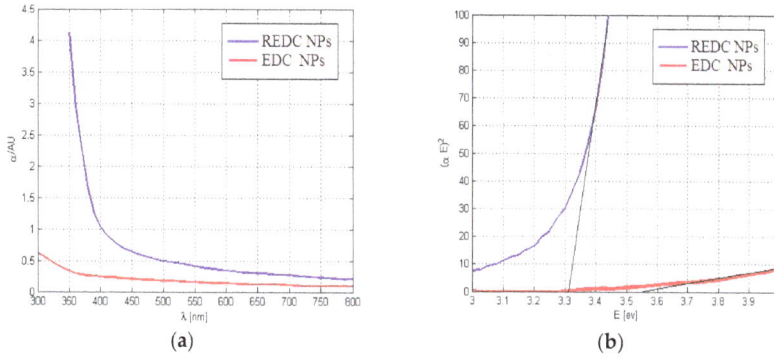

Figure 1. (**a**) Absorbance dispersion curves for reduced nanoparticles (REDC NPs) annealed at 700 °C and the non-reduced nanoparticles (EDC NPs); (**b**) the corresponding direct band gap of both REDC NPs and EDC NPs.

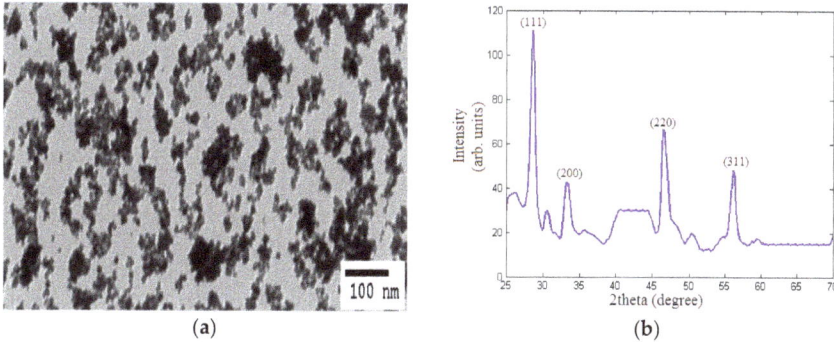

Figure 2. (**a**) TEM image and (**b**) XRD pattern of REDC NPs at annealing temperature of 700 °C.

Under the simultaneous emission of both near-UV ($\lambda = 430$ nm) and IR (780 nm) excitations, the dominant visible emission from the EDC NPs is centered around 520 nm with a relatively smaller-peak emission at 670 nm, as shown in Figure 3. This emission peak is including both contributions; the down-conversion one which involves the radiative relaxation of the 5d–4f transition of an excited Ce^{3+} ion in Ce_2O_3 resulting in the broadband emission of the green wavelength [19]. As the synthesized REDC NPs that contain some fraction of Ce_2O_3 are illuminated with near-UV light, then some fraction of the valence band electrons are excited to an oxygen vacancy defect state located within the CeO_2 band gap. Regarding the second contribution of the up-conversion process, erbium ions form stable complexes with oxygen in the ceria host during the annealing, and the crystalline structure of the nanoparticle improves, both of which increase the efficiency of

Er^{3+} ions to behave as optically active centers for up-conversion emissions with the mutual contribution of green light in addition to the low red emission [15,19]. In addition, the conductivity of the REDC NPs is measured to be 77 μS/cm, which is found experimentally to be ~22% higher than the conductivity of normal EDC NPs, 63 μS/cm.

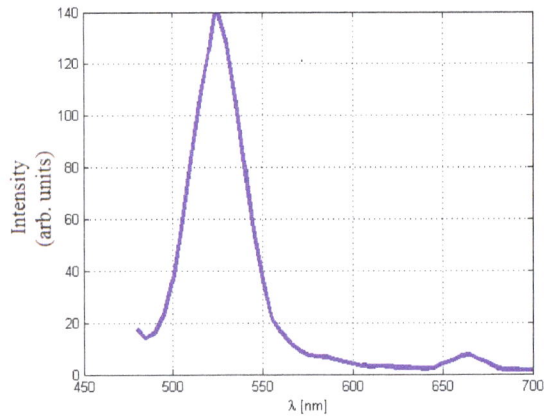

Figure 3. Emission spectrum of REDC NPs under simultaneous excitations of both near-UV (430 nm) and IR (780 nm) excitations.

The surface profile of the coated cell is shown in Figure 4, with focus on the region between the electrode and the coated edge. It could be observed that the mean thickness of the coating is around 20 nm with quite a non-uniform distribution of the coating, as shown in Figure 4b regarding the intensity imaging, which may be due to the spin coating technique itself. This coating technique could be considered as a trade-off between surface uniformity and simplicity. However, other coating techniques may lead us to miss the conductivity of the nanostructures due to missing oxygen vacancies with the conversion of Ce^{3+} to Ce^{+4}.

2.2. Coated Solar Cell Characterization

As investigated in the previous sections, coating the back side of a silicon solar cell with REDC NPs has the advantages of improving multi-optical conversions, leading to the conversion of some UV and IR wavelengths that solar cells cannot absorb to visible light wavelengths which can be absorbed. Figure 5a,b show the improvement in P–V and I–V curves, respectively, after coating the cell with REDC NPs. The promising comparison between coated and uncoated cells was shown in Table 1, and it clearly shows that power conversion efficiency (PEC) has been improved from 15.1% to 16.7%, which is about a 10.8% improvement of cell efficiency due to coating compared to uncoated cells. As can be noticed from

Table 1, there is a relatively high improvement of short circuit current ($I_{s.c}$) with the effect of our synthesized nanoparticle coating compared to both quite stable open circuit voltage ($V_{o.c}$) and fill factor (F.F). Overall, the increase in the current, and consequently the power, could be explained due to the increase in the rate of photoelectrons, whether through a higher generation rate due to optical conversions and/or the better mobility due to a conductive nanostructure coating.

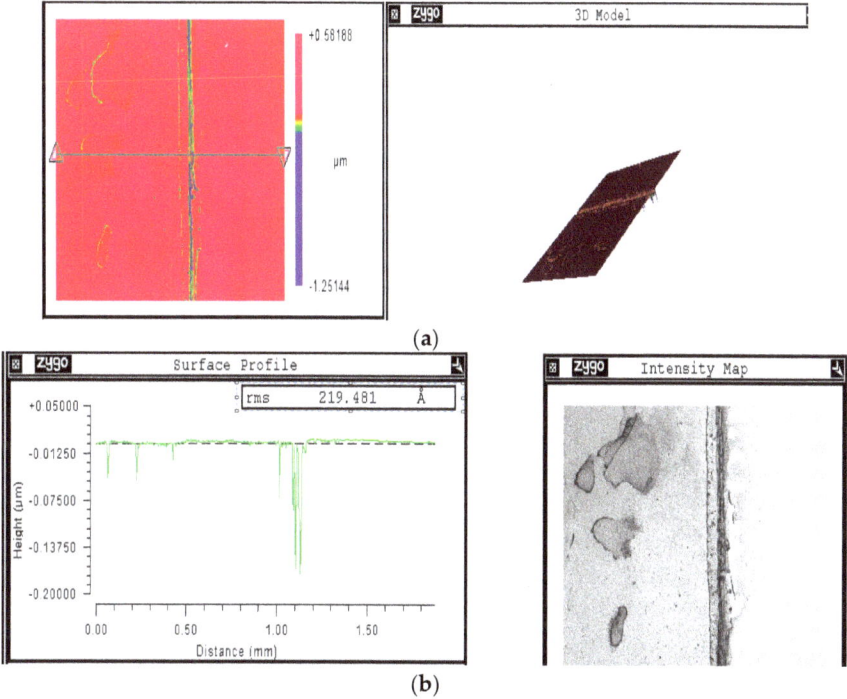

(a)

(b)

Figure 4. (a) Surface profile of coated cell at the edge between coating and the electrode and (b) the profile distribution with the intensity map.

Table 1. Comparison between coated and un-coated cells.

Condition	$V_{o.c}$	$I_{s.c}$	F.F	η %
Uncoated	0.5155	0.1537	0.6301	15.1075
coated cell	0.5095	0.1718	0.6322	16.7452

179

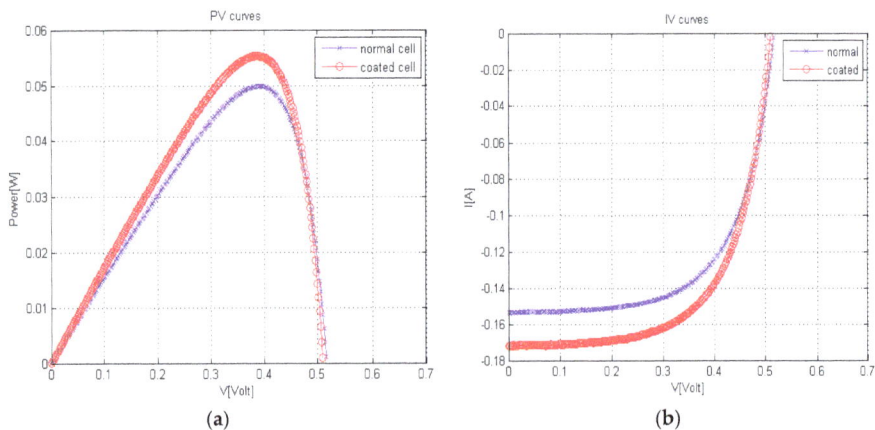

Figure 5. (a) P–V curve and (b) I–V curve of silicon solar cells in both the uncoated (normal) case and those coated with REDC NPs.

Beside the advantage of multi-optical conversions of REDC NPs, these nanoparticles have the ability to improve the electrical conductivity of the generated photoelectrons of solar cells through the great number of formed O-vacancies. Then, we aimed to simulate Si solar cells before and after the REDC NP layer coating through studying the normalized generation rate and field distribution. Figure 6 shows the difference in generation rate curves, and the surface electric field distributions are shown in Figure 7a,b. A simulation model has been built in a two-dimensional (2D), semiconductor module. This model deals with REDC NPs as it is a conductive layer with a band gap E_g = 3.31 eV, room temperature conductivity σ = 77 × 10^{-6} S/cm, and electron mobility μ_e = 2.8 × 10^{-7} cm^2/V·s [20,21]. From Figure 6, it has been proved that a REDC NP coated cell has a little bit of improvement in the generation rate curve. The difference between the maximum of the curves before and after the NP coating layer is calculated to be about 0.408%. That gives an indication that the conductivity impact of the coating nanoparticles has a major impact in the solar cell's efficiency increase rather than the optical conversions.

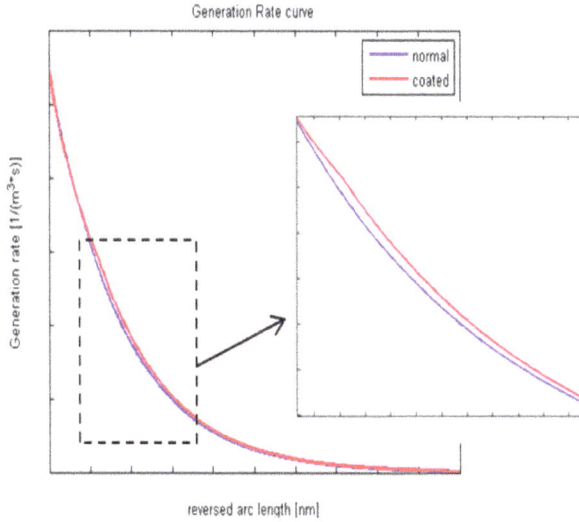

Figure 6. Normalized generation rate of silicon cells with/without REDC NP coating layer.

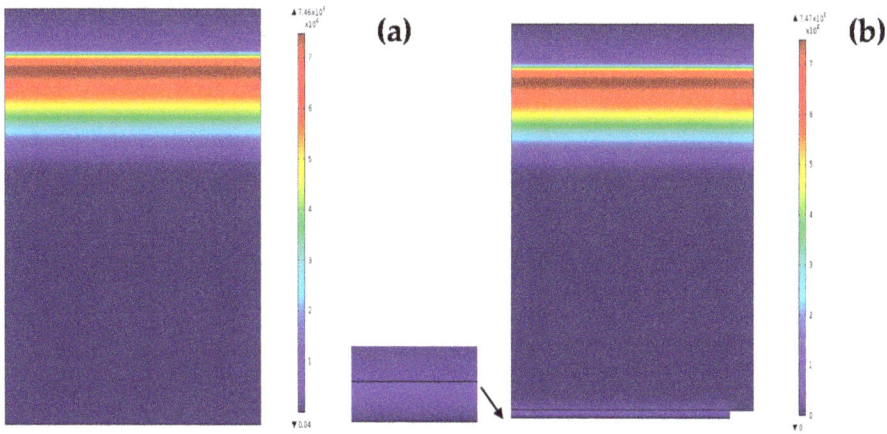

Figure 7. Electric field distribution (**a**) before and (**b**) after REDC NP coating.

Electric field distribution before and after adding a REDC NP layer is shown in Figure 6a,b, respectively. There is some concentrated electric field between the solar cell and the REDC NP layer, with a slight difference in the electric field maximum value which was about 0.1340%. That could give a conclusion that the added layer of REDC NPs could have a slight optical impact in concentrating the electric fields inside the solar cells, in addition to the interface region between the cell and the coating layer. This confirms the mutual impact of the improved optical conversions

181

and conductivity due to the REDC NP coating layer, with the dominant effect of the conductivity due to the hopping mechanism of the formed oxygen vacancies inside REDC NPs.

3. Experimental Section

3.1. Nanoparticles Synthesis

Reduced erbium-doped ceria nanoparticles have been synthesized using the chemical precipitation technique which is a relatively simple and inexpensive synthesis process [22]. Cerium (III) chloride (heptahydrate, 99.9%, Sigma-Aldrich Chemicals, St. Louis, MO, USA) of weight 0.485 g and erbium (III) chloride (heptahydrate, 99.9%, Sigma-Aldrich Chemicals, St. Louis, MO, USA) (0.015 g) are dissolved in de-ionized (DI) water (40 mL) to obtain a 3% weight ratio of erbium to cerium in the synthesized nanoparticles. This weight ratio is selected after a study by the authors of different weight ratios of erbium-doped ceria nanoparticles, synthesized using the same process, in which it was found that the optimal concentration of erbium in ceria for up-conversion is 3 wt % for the optimum improvement of solar cell efficiency. The solution is stirred constantly at 500 rpm in a water bath, while the temperature of the water bath is raised to 60 °C, and ammonia (1.6 mL) is then added to the solution. The solution is kept at 60 °C for 2 h and, then, the solution is stirred for another 22 h at room temperature. Then, the wet powder is dried, after being washed using ethanol, on a hot plate for 20 min. The thermal annealing of the dried nanoparticles is performed in a tube furnace (CM Furnace, Model 1730-20HT, Bloomfield, NJ, USA) with an atmosphere of hydrogen and nitrogen gases that are injected into the furnace at flow rates equal to 10 standard cubic feet per minute (scfm) for two hours at temperatures of 700 °C. The gases during the annealing assist with the reduction of the cerium ions from the Ce^{4+} to Ce^{3+} ionization states and the creation of the oxygen vacancies [23–26], while the thermal energy available during the high temperature anneal promotes the formation of the molecular energy levels of erbium inside the ceria host [15].

3.2. Characterizations of Nanoparticles

The optical absorption is measured using a dual-beam UV-Vis-NIR spectrometer (UV-3101PC Shimadzu, Tokyo, Japan). After the annealing procedure, a solution of nanoparticles is prepared with a concentration of 0.02 mg of nanoparticles in 10 mL of DI water. The colloidal solution is illuminated with both near-UV and near infra-red (NIR) excitations in an experimental apparatus that was designed to measure the down- and up-conversion process, as described in Figure 8. The fluorescence spectroscopy system consists of two excitation sources. The first one, the near UV excitation, is a Xenon lamp coupled to a monochromator, (Cornerstone 260, Newport, Irvine, CA, USA). The light that exits the monochromator (λ_{exc} = 430 nm) is focused on to the colloidal solution. The second one is an IR laser module of 780 nm. Both down- and up-conversion are detected using a second monochromator (Cornerstone 260, Newport, Irvine, CA, USA), positioned at a 90° angle to the first monochromator. The monochromator is scanned over the visible wavelength region and the fluorescence signal is detected by the photomultiplier tube (PMT 77340, Newport, Irvine, CA, USA), located at the exit port of the second monochromator. Then, the visible fluorescent emission is monitored using a power meter (2935C, Newport, Irvine, CA, USA).

Transmission electron microscope (TEM) (JEOL 1400, Peabody, MA, USA), is used to image the synthesized REDC NPs. The mean diameter of the nanoparticles is calculated using ImageJ software through Gaussian distribution of many size measurements. The operating parameters of the XRD (PANalytical X'Pert PRO, Amestrdam, The Netherlands), are 45 KV, 40 A and Cu Kα radiation (λ = 0.15406 nm). The conductivity of the solution of the synthesized nanoparticles is measured by A500 Orion meter (Thermo scientific, Tech Park, Singapore).

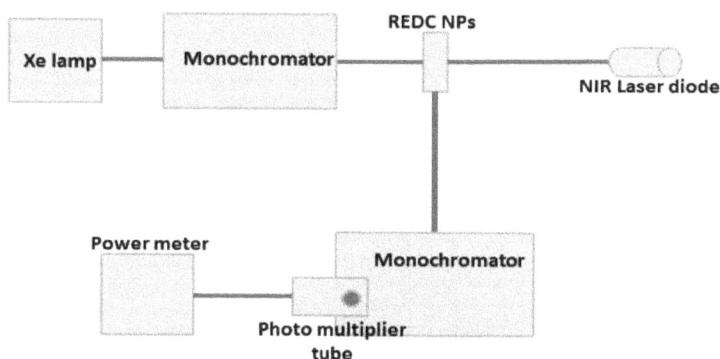

Figure 8. Up- and down-conversion fluorescence setup.

3.3. Coating Procedure

The synthesized nanoparticles are coated on the back sides of polycrystalline solar cells (2 inches × 2 inches) ordered from Solar Winds Inc., Austin, TX, USA. Coating has been operated using spin coater at 1500 rpm for a minute. Before coating, the electrodes on the backside of the cell have been covered through a scotch and released after the coating. That could avoid the direct contact between the nanoparticles and the metallic electrodes. The surface profile is detected using 3D optical surface profiler ZeGage (Zygo, Middlefield, CT, USA), with concentrating on the edge between coated cell and non-coated electrode to detect the thickness.

3.4. Solar Cell Characterization

The nanoparticle-coated and uncoated solar cells have been analyzed using a designed I–V characterization setup. The irradiance generated from a Xenon lamp (Oriel 100W, Irvine, CA, USA) followed by air mass (Newport AM1.5G, Irvine, CA, USA) is exposed to the coated/uncoated solar cells. Then, the different values of I and V are measured using Source meter 2400-C source meter (Keithley, Cleveland, OH, USA), with sweeping parameters as voltage range from -1 to $+1$ V through 1000 measuring points with 50 ms stoppage time per reading. Through the extracted *I*-values corresponding to the swept V-values, both I–V and P–V curves are drawn. From the I–V curve, some parameters could be measured such as filling factor, $V_{o.c}$, $I_{s.c}$ and the optical efficiency. Using COMSOL Multiphysics software (COMSOL Inc., Burlington, MA, USA), generation rate and E-field distribution are analyzed with and without the nanoparticle layer on silicon solar cell.

4. Conclusions

This paper introduces a novel study of using reduced erbium-doped ceria nanoparticles (REDC NPs) as a coating layer on silicon solar cells. The presented work shows full optical characterization of the synthesized nanoparticles. The experimental results show the visible fluorescence emitted under both excitations of NIR and near UV. In addition, the results of the band gap and fluorescence confirm the formation of Ce^{3+} trap states which are associated with the formation of charged oxygen vacancies. That could increase the conductivity for any photo-generated electrons in the host NPs. When depositing REDC NPs on the back sides of solar cells, a promising improvement in the solar cell efficiency has been observed from 15% to 16.5% due to the mutual impact of improved electric conductivity and multi-optical conversions. In addition, the generation rate and maximum electric fields formed in the solar cells have been slightly improved due to the coating.

Acknowledgments: This work was funded in part by a NSF STTR Phase I grant with MW Photonics (award 0930364). Also, some the authors are supported through Virginia Tech Middle East and North Africa (VT-MENA) program and center of SmartCI research center in Alexandria University. The authors appreciate the support of both Ibrahim Hassounah and Michael Ellis' lab in ICTAS, Virginia Polytechnic Institute State University (Virginia Tech), in the process of annealing the synthesized nanoparticles. Also, the authors appreciate the support of Don Leber; manager of the Micron Technology Semiconductor Processing Laboratory at Virginia Tech.

Author Contributions: Nader Shehata was responsible for preparation and characterization of the nanoparticles. Michael Clavel was responsible for coating procedure and solar cell experimental setup under the supervision of both Nader Shehata and Kathleen Meehan. Effat Samir was the main person for solar cell analysis under the supervision of Mohammed Salah. Soha Gaballah did the TEM image of the nanoparticles.

Conflicts of Interest: The authors declare no conflict of interest.

References

1. Maruyama, T.; Shinyashiki, Y.; Osako, S. Energy conversion efficiency of solar cells coated with fluorescent coloring agent. *Sol. Energy Mater. Sol. Cells* **1998**, *56*, 1–6.

2. Shan, G.D.; Demopoulos, G.P. Near-infrared sunlight harvesting in dye-sensitized solar cells via the insertion of an upconverter-TiO_2 nanocomposite layer. *Adv. Mater.* **2010**, *22*, 4373–4377.

3. Tsunekawa, S.; Fukuda, T.; Kasuya, A. Blue shift in ultraviolet absorption spectra of monodisperse CeO_{2-x} nanoparticles. *J. Appl. Phys.* **2000**, *87*, 1318–1321.

4. Oh, M.; Nho, J.; Cho, S.; Lee, J.; Sing, R. Polishing behaviors of ceria abrasives on silicon dioxide and silicon nitride CMP. *Powder Technol.* **2011**, *206*, 239–245.

5. Das, M.; Patil, S.; Bhargava, N.; Kang, J.; Riedel, L.; Seal, S.; Hickman, J. Auto-catalytic ceria nanoparticles offer neuroprotection to adult rat spinal cord neurons. *Biomaterials* **2007**, *28*, 1918–1925.

6. Steel, B.; Heinzel, B. Materials for fuel-cell technologies. *Nature* **2001**, *414*, 345–352.

7. Chen, J.; Chang, W.; Huang, C.; Sun, K. Biomimetic nanostructured antireflection coating and its application on crystalline silicon solar cells. *Opt. Express* **2011**, *18*, 14411–14419.

8. Basu, T.; Ray, M.; Ratan, N.; Pramanick, A.; Hossain, S. Performance enhancement of crystalline silicon solar cells by coating with luminescent silicon nanostructures. *J. Electron. Mater.* **2013**, *42*, 403–409.

9. Pinna, A.; Figus, C.; Lasio, B.; Piccinini, M.; Malfatti, L.; Innocenzi, P. Release of ceria nanoparticles grafted on hybrid organic–inorganic films for biomedical application. *ACS Appl. Mater. Interfaces* **2012**, *4*, 3916–3922.

10. Pinna, A.; Lasio, B.; Piccinini, M.; Marmiroli, B.; Amenitsch, H.; Falcaro, P.; Tokudome, Y.; Malfatti, L.; Innocenzi, P. Combining top-down and bottom-up routes for fabrication of mesoporous titania films containing ceria nanoparticles for free radical scavenging. *ACS Appl. Mater. Interfaces* **2013**, *5*, 3168–3175.

11. Pinna, A.; Barbara, B.; Lasio, B.; Malfatti, L. Engineering the surface of hybrid organic-inorganic films with orthogonal grafting of oxide nanoparticles. *J. Nanopart. Res.* **2014**, *16*, 2463–2466.

12. Pankove, P. *Optical Processes in Semiconductors*; Dover Publications Inc.: New York, NY, USA, 1971.

13. Shehata, N.; Meehan, K.; Leber, D. Fluorescence quenching in ceria nanoparticles: A dissolved oxygen molecular probe with a relatively temperature insensitive Stern-Volmer constant up to 50 °C. *J. Nanophotonics* **2012**, *6*.

14. Dhannia, T.; Jayalekshmi, S.; Kumar, M.; Rao, T.; Bose, A. Effect of aluminium doping and annealing on structural and optical properties of cerium oxide nanocrystals. *J. Phys. Chem. Solids* **2009**, *70*, 1443–1447.

15. Lawrence, N.; Jiang, K.; Cheung, C.L. Formation of a porous cerium oxide membrane by anodization. *Chem. Commun.* **2011**, *47*, 2703–2705.

16. Shehata, N.; Meehan, K.; Hassounah, I.; Hudait, M.; Jain, N.; Clavel, M.; Elhelw, S.; Madi, N. Reduced erbium-doped ceria nanoparticles: One nano-host applicable for simultaneous optical down- and up-conversions. *Nanoscale Res. Lett.* **2014**, *9*.

17. Basu, S.; Devi, S.; Maiti, S. Synthesis and properties of nanocrystalline ceria powders. *J. Mater. Res.* **2004**, *19*, 3162–3171.

18. Guo, H. Green and red upconversion luminescence in CeO_2:Er^{3+} powders produced by 785 nm laser. *J. Solid State Chem.* **2007**, *180*, 127–131.

19. Liu, T.; Hon, M.; Teoh, L.G. Structure and optical properties of CeO_2 nanoparticles synthesized by precipitation. *J. Electron. Mater.* **2013**, *42*, 2536–2541.

20. Lappalainen, J.; Tuller, H.; Lantto, V. Electronic conductivity and dielectric properties of nanocrystalline CeO_2 films. *J. Electroceram.* **2004**, *13*, 129–133.

21. Qiu, L.; Liu, F.; Zhao, L.; Ma, Y.; Ya, J. Comparative XPS study of surface reduction for nanocrystalline and microcrystalline ceria powder. *Appl. Surf. Sci.* **2006**, *25*, 4931–4935.

22. Chen, H.; Chang, H. Homogeneous precipitation of cerium dioxide nanoparticles in alcohol/water mixed solvents. *Colloids Surf. A* **2004**, *242*, 61–69.

23. Shehata, N.; Meehan, K.; Hudait, M.; Jain, N. Control of oxygen vacancies and Ce^{+3} concentrations in doped ceria nanoparticles via the selection of lanthanide element. *J. Nanopart. Res.* **2012**, *14*, 1173–1183.

24. Chui, C.O.; Kim, H.; McIntyre, P.C.; Saraswat, K.C. Atomic layer deposition of high-κ dielectric for germanium MOS applications—Substrate surface preparation. *IEEE Electron. Device Lett.* **2004**, *25*, 274–276.

25. Trovarelli, A. *Catalysis by Ceria and Related Materials*; Imperial College Press: London, UK, 2005.

26. Shehata, N.; Meehan, K.; Leber, D. Study of fluorescence quenching in aluminum-doped ceria nanoparticles: Potential molecular probe for dissolved oxygen. *J. Fluoresc.* **2013**, *23*, 527–532.

Towards InAs/InGaAs/GaAs Quantum Dot Solar Cells Directly Grown on Si Substrate

Bilel Azeza, Mohamed Helmi Hadj Alouane, Bouraoui Ilahi, Gilles Patriarche, Larbi Sfaxi, Afif Fouzri, Hassen Maaref and Ridha M'ghaieth

Abstract: This paper reports on an initial assessment of the direct growth of In(Ga)As/GaAs quantum dots (QDs) solar cells on nanostructured surface Si substrate by molecular beam epitaxy (MBE). The effect of inserting 40 InAs/InGaAs/GaAs QDs layers in the intrinsic region of the heterojunction pin-GaAs/n^+-Si was evaluated using photocurrent spectroscopy in comparison with pin-GaAs/n^+-Si and pin-GaAs/GaAs without QDs. The results reveal the clear contribution of the QDs layers to the improvement of the spectral response up to 1200 nm. The novel structure has been studied by X ray diffraction (XRD), photoluminescence spectroscopy (PL) and transmission electron microscopy (TEM). These results provide considerable insights into low cost III-V material-based solar cells.

Reprinted from *Materials*. Cite as: Azeza, B.; Alouane, M.H.H.; Ilahi, B.; Patriarche, G.; Sfaxi, L.; Fouzri, A.; Maaref, H.; M'ghaieth, R. Towards InAs/InGaAs/GaAs Quantum Dot Solar Cells Directly Grown on Si Substrate. *Materials* **2015**, *8*, 4544–4552.

1. Introduction

Recent attempts have been made to increase solar cell efficiency by exploiting the below band gap photon energies. This approach is often termed as impurity band solar cells or intermediate band solar cells (IBSC) [1–4]. This type of structure demonstrates an enhancement of the spectral response towards lower photon energies with promising possibilities to improve the solar cell's efficiency up to 63% according to theoretical expectation [5–8]. However, the main problem encountered for III-V element-based solar cells mainly lies in the development of large surface photovoltaic structures due to their high cost, making the employment of low cost substrates highly desirable. Accordingly, the epitaxial growth of the GaAs layer on Si substrate has attracted considerable attention owing to their large area availability, low cost and high mechanical strength [9–14]. Promising InAs/GaAs QD-based optoelectronic devices, directly grown on Si substrate, have already been reported [15–20]. In the meantime, with the exception of the employment of the bonding technique [21], the feasibility of the direct growth of InAs/GaAs QDs solar cells on Si substrate has not yet been explored.

In this context, this paper reports, for the first time, on the effect of inserting InAs/InGaAs/GaAs multiple QDs layers within the pin-GaAs structure directly

187

deposited by MBE on nanostructured Si substrate for solar cell applications. The results may open up new perspectives on the development of low cost, high efficiency III-V-based solar cells on Si substrate.

2. Results and Discussion

2.1. Growth Process

InAs/GaAs QDs with $In_{0.13}Ga_{0.87}As$ strain reducing layer were incorporated within the intrinsic region of a pin-GaAs/n^+-Si using Stranski-Kranstanow growth mode by molecular beam epitaxy. The use of InGaAs as a strain reducing layer is believed both to reduce the compressive stress acting on the InAs QDs by the GaAs matrix, and to reduce indium out-diffusion from the InAs QDs [22–24]. The choice of relatively low indium composition is expected to avoid excessive In-Ga phase separation that alters the optical and structural properties of the InAs QDs and the surrounding material [25]. The number of absorbed photons is proportional to the number of effective QDs in the solar cell active region. For the InAs/GaAs system, the QDs' aerial density in a single layer is around 10^{10} dots/cm^2 on GaAs substrate [26]. Such a value is too small to account for the improvement of the spectral response. Increasing the effective number of QDs is possible by vertical stacking of QD layers. However, the total number of vertically stacked layers is limited by the onset required for the relaxation of the accumulated strain by generation of stacking faults and dislocations. In the present study, we have employed 40 QD layers ensuring a compromise between the increase of the effective number of QDs and the overall sample's structural properties. The pin diode structure has been directly fabricated on nanostructured n^+-Si substrate.

The preparation of the nanostructured n+-type silicon's substrate surface has been performed at room temperature by the formation and subsequent dissolution of a porous layer. The electrolyte used to fabricate a 5 μm-thick porous silicon layer consists of a mixture of hydrofluoric acid HF ([HF] = 36%) and ethanol (HF:C2H5OH) in a volumetric proportion of 1:1. The porous layer was formed by anodizing the Si substrate in this electrolyte under a current density of 3 mA·cm^{-2}. The sample was then etched in NaOH solution to break up the porous silicon layer and produce the structuration of the surface. Indeed, after the chemical dissolution of silicon skeleton, the rugged surface will be exposed to beam epitaxy. Additional details concerning the process as well as the morphological properties of the nanostructured Si surface and its impact on the quality of GaAs material grown on such Si surface can be found elsewhere [27].

After surface preparation, a cleaning and out gassing process of the silicon substrate was performed under vacuum condition in an introductory chamber with a rest pressure of 10^{-9} Torr at high temperature (760 °C), to remove the native oxide and other volatile compounds prior to the GaAs deposition. The growth began by depositing 0.25 µm n+ doped GaAs layer at 530 °C followed by 1 µm of n doped GaAs layer at 580 °C. Forty layers of 0.7 nm nominal thickness InAs QDs capped first by 5 nm $In_{0.13}Ga_{0.87}As$ and then 7 nm GaAs spacer layer were subsequently deposited at 500 °C. Finally, 0.5 µm p doped GaAs was grown at 580 °C followed by 0.1 µm of p$^+$ doped GaAs layer at 530 °C. The growth rate was: 0.24 Ås^{-1} for the InAs, 2.22 Ås^{-1} for the InGaAs and 1.98 Ås^{-1} for GaAs.

Two reference pin-GaAs diodes without QDs have also been fabricated under the same conditions either on nanostructured Si substrate and GaAs substrate. A schematic presentation of the pin-GaAs diode on Si substrate with and without QDs is given in Figure 1.

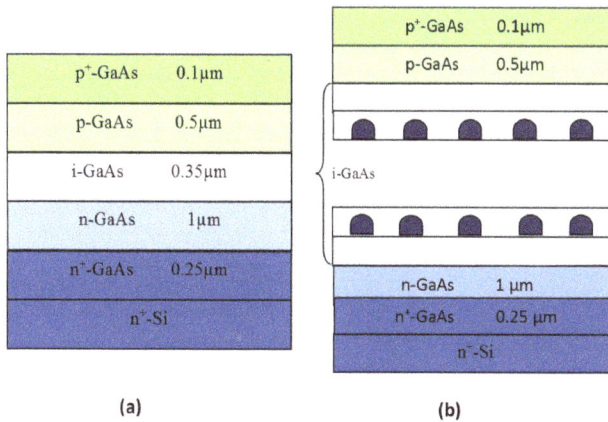

Figure 1. Schematic presentation of the investigated samples. (**a**) pin-GaAs/Si; (**b**) pin-GaAs/Si containing 40 QD layers.

During the growth process, the surface morphology was *in-situ* monitored by reflection high-energy electron diffraction (RHEED). As shown by Figure 2, the RHEED pattern changed from streaky (Figure 2a) during the GaAs deposition, which is characteristic of 2D growth mode to a spotty pattern (Figure 2b) after the deposition of InAs material. The observed changes in the diffraction pattern represent the transition from 2D to 3D growth mode, testifying the QDs' formation.

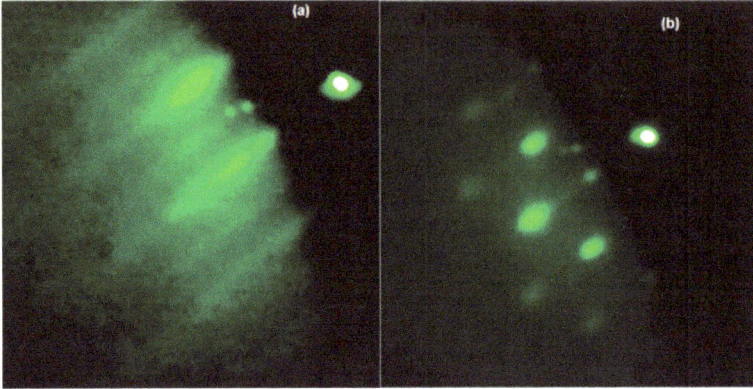

Figure 2. RHEED diffraction pattern. (**a**) During the growth of GaAs; (**b**) after the deposition of the InAs QDs.

2.2. Material Characterization

Figure 3 illustrates the $\omega/2\theta$, reflection peaks of pin-GaAs/n$^+$-Si with and without InAs/InGaAs multistaked QDs.

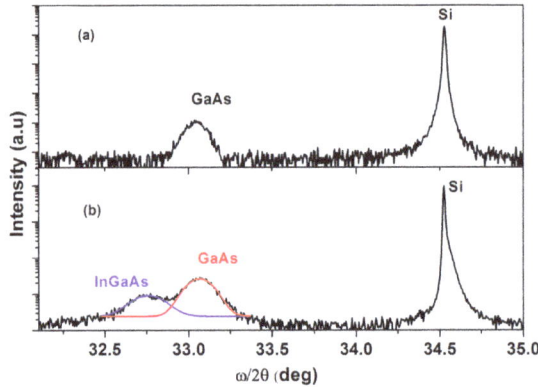

Figure 3. The $\omega/2\theta$ of reflection peak of: (**a**) pin-GaAs/ Si; (**b**) pin-GaAs/Si with InAs/InGaAs multilayer QDs.

The spectra from the pin-GaAs (a) and pin-GaAs with QDs (b) grown on Si substrate reveal the presence of two peaks centered on $\theta = 34.52°$ and $\theta = 33.06°$ and attributed respectively to the silicon substrate and to the GaAs layer. Additionally, a third peak appear at $\theta = 32.75°$ in the XRD spectra of the structure containing the QDs. This peak can be attributed to the InAs/InGaAs multilayer with a nominal indium composition x_m equal to the average of indium compositions in all layers ($x_m = 13.27\%$ estimated by HRXRD). For the InAs/GaAs multistaked QDs grown on

190

GaAs substrate, the HRXRD spectra show the appearance of other peaks appointed satellite peaks, due to the periodicity introduced by the bilayers repetition and the angular period of this peak is related to the thickness of the bilayer [28,29]. In our case, the absence of satellite peaks could be explained by the existence of defects produced in the interfaces layers. Indeed, as shown by the Figure 4, the cross section transmission electron microscopy image unambiguously shows that the GaAs buffer layer was not sufficiently smooth. The surface roughness greatly influenced the multiple layer QDs, resulting in distorted layers. Consequently, the grown InAs/InGaAs QDs display a non-uniform thickness which in turn provokes plastic strain relaxation via defects and threading dislocations.

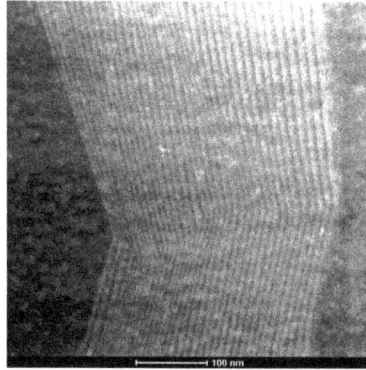

Figure 4. Cross section TEM image of the InAs/InGaAs multilayer QDs.

Additional details can be given by PL characterization. Figure 5 shows the 11 K PL spectra of the pin-GaAs/n^+-Si structures with and without QDs. A peak centered at 842 nm appear in both structures and are attributed to the GaAs' emission. The red shift of GaAs emission peak is a consequence of the lattice mismatch between GaAs and Si, the polar/nonpolar character and of the strong tensile stress, since the thermal expansion coefficient of GaAs is about twice that of the silicon value. The low intensity of these peaks is directly linked to the subsistence of non-radiative recombination channels due to the defects in the structure.

For the structure containing multiple layer QDs, the PL measurement reveals a broad band centered at 1100 nm. This band is likely to arise from the luminescence of the InAs QDs. Although this result confirms the formation of InAs/GaAs QDs, the broadening of the PL band with relatively weak intensity confirms that the QDs structural properties are altered.

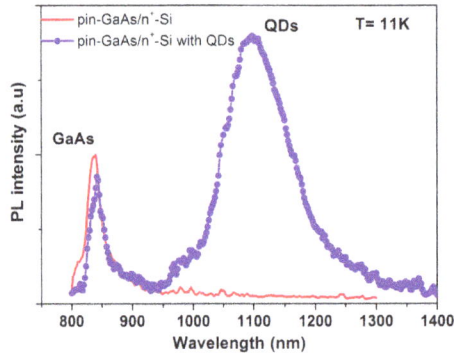

Figure 5. PL spectra recorded at 11 k from the pin-GaAs/Si structure (red line) and from pin-GaAs/Si with InAs/InGaAs multilayer QDs (blue line).

To further assess the impact of introducing the InAs QDs within the pin-GaAs/Si we have performed the spectral response measurements from samples with and without QDs. The results are shown by Figure 6. The photo-response obtained from pin-GaAs/n$^+$-Si without QDs for the high energy photons (beyond the GaAs band gap) produces the same range of photo-response obtained by the reference cell grown on GaAs substrate. This assures that the photocarriers collected by the structure are mainly created by the pin-GaAs prepared on the Si substrate. However, for lower energy photons, the spectral response of the reference cell drops abruptly at 868 nm corresponding to the band gap energy of GaAs (1.42 eV). In the meantime, the photo-response from pin-GaAs/n$^+$-Si recovers a proportion of the below GaAs band gap photons to an extent of up to 1200 nm. The observed enhancement is due to photocarriers generated by silicon substrate.

A more pronounced improvement in the photo-response at long wavelengths is observed for the structure containing QDs. This improvement is due to the absorption of photons below the band gap energy of GaAs by InAs QDs.

Although the structural properties of the multiple QDs appear rather to be degraded principally as a consequence of the initial surface roughness, the optical and electric properties of pin-GaAs/n+-Si with InAs QDs show that the InAs QDs have been formed and successively contribute to the electron–hole pair creations in the below band gap energy range which increased the photocarrier collections [30,31]. This initial assessment provides evidence of the potential of our proposed yielding structures for the fabrication of future novel, low cost, high performance solar cells.

At this time, no contact grid coatings were applied and the electrical contact was basically made with indium-zinc alloys pads on the front surface. *In-situ* and *ex-situ* optimization of the solar cell fabrication is in progress, a necessary step to obtain significant values from the active solar cell parameters.

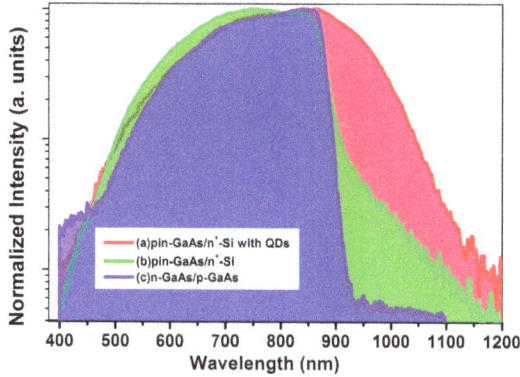

Figure 6. Spectral response of (**a**) pin-GaAs/ n$^+$-Si with QDs (red); (**b**) pin-GaAs/n$^+$-Si without QDs (green); (**c**) reference pin-GsAs on GaAs substrate (blue).

3. Experimental Section

The HRXRD experiments were performed with D8 DISCOVER Bruker Axs Diffractometer (BRUKER, Karlsruhe, Germany) with CuKα1 radiation (λ CuKα = 1.5406Å) for $\omega/2\theta$ values in the range of 32°–35° to investigate the structural properties of GaAs layer grown on nanostructured Si substrate.

The PL measurements have been done at 11 K, and the samples mounted in a closed cycle He cryostat, were excited with the 514.5 nm line of an Ar$^+$ laser (Spectra-Physics, Santa Clara, CA, USA) while the spectra were collected using a thermoelectrically cooled InGaAs photodetector (Oriel, Stratford, CT, USA) using a conventional lock-in technique.

The cross section transmission electron microscopy image was performed using a TEM/STEM Cs-corrected JEOL 2200 FS (JEOL, Peabody, MA, USA) operated at 200 kV.

The spectral response measurements aim to evaluate the electrical current photogenerated in our samples. The spectral response is measured using a 100 W tungsten halogen lamp (Newport, Santa Clara, CA, USA), CVI CM110 1/8 m monochromator (Spectral Product, Cvijovica Dolina, CA, USA) and a lock-in amplifier connected to a chopper (at 172 Hz) placed at the outlet of the source monochromator.

4. Conclusions

InAs/GaAs QD-based pin GaAs solar cells directly grown on silicon substrate have been demonstrated for the first time by using the nanostructured surface as a buffer layer. This initial assessment shows the formation of InAs nanostructure, with an emission wavelength of around 1100 nm. The insertion of multiple layer QDs within a pin GaAs grown on Si substrate has been found to improve the spectral response to 1200 nm, despite imperfect structural properties. These results hold great promise for future demonstration of high efficiency IBSCs on Si substrate via heteroepitaxy using nanostructured surfaces.

Acknowledgments: The third author (Bouraoui Ilahi) would like to extend his sincere appreciation to the Deanship of Scientific Research at King Saud University for funding this Research Group NO.: RG-1436-014.

Author Contributions: Bilel Azeza and Ridha M'ghaieth did the photocurrent measurements. Mohamed Helmi Hadj Alouane and Bouraoui Ilahi did the PL measurements. Larbi Sfaxi and Hassen Maaref did the samples growth. Gilles Patriarche did the TEM image and Afif Fouzri did the XRD measurements. All the authors have contributed to the interpretation of the results and revision of the paper written by Bilel Azeza.

Conflicts of Interest: The authors declare no conflict of interest.

References

1. Luque, A.; Stanley, C. Understanding intermediate-band solar cells. *Nat. Photonics* **2012**, *6*, 146–152.
2. Luque, A.; Marti, A. Increasing the efficiency of ideal solar cells by photon induced transitions at intermediate levels. *Phys. Rev. Lett.* **1997**, *78*, 5014–5017.
3. Marti, A.; Antolin, E.; Stanley, R.C.; Farmer, C.D.; Lopez, N.; Diaz, P.; Canovas, E.; Linares, G.P.; Luque, A. Production of photocurrent due to intermediate-to-conduction-band transitions: A demonstration of a key operating principle of the intermediate-band solar cell. *Phys. Rev. Lett.* **2006**, *97*.
4. Marrón, D.F.; Artacho, I.; Stanley, R.C.; Steer, M.; Kaizu, T.; Shoji, Y.; Ahsan, N.; Okada, Y.; Barrigón, E.; Rey-Stolle, I.; *et al.* Application of photoreflectance to advanced multilayer structures for photovoltaics. *Mater. Sci. Eng. B* **2013**, *178*, 599–608.
5. Wu, J.; Makableh, Y.M.F.; Vasan, R.; Manasreh, M.O.; Liang, B.; Reyner, C.J.; Huffaker, D.L. Strong interband transitions in InAs quantum dots solar cell. *Appl. Phys. Lett.* **2012**, *100*.
6. Bailey, C.G.; Forbes, D.V.; Raffaelle, R.P.; Hubbard, S.M. Near 1 V open circuit voltage InAs/GaAs quantum dot solar cells. *Appl. Phys. Lett.* **2011**, *98*.
7. Guimard, D.; Morihara, R.; Bordel, D.; Tanabe, K.; Wakayama, Y.; Nishioka, M.; Arakawa, Y. Fabrication of InAs/GaAs quantum dot solar cells with enhanced photocurrent and without degradation of open circuit voltage. *Appl. Phys. Lett.* **2010**, *96*.
8. Linares, P.G.; Marti, A.; Antoli, E.; Farmer, C.D.; Ramiro, I.; Stanley, C.R.; Luque, A. Voltage recovery in intermediate band solar cells. *Sol Energy Mater. Sol. Cells* **2012**, *98*, 240–244.
9. Soga, T.; Jimbo, T.; Arokiaraj, J.; Umeno, M. Growth of stress-released GaAs on GaAs/Si structure by metalorganic chemical vapor deposition. *Appl. Phys. Lett.* **2000**, *77*.
10. Azeza, B.; Ezzedini, M.; Zaaboub, Z.; M'ghaieth, R.; Sfaxi, L.; Hassen, F.; Maaref, H. Impact of rough silicon buffer layer on electronic quality of GaAs grown on Si substrate. *Curr. Appl. Phys.* **2012**, *12*, 1256–1258.

11. Vanamu, G.; Datye, A.K.; Dawson, R.; Zaidi, S.H. Growth of high-quality GaAs on Ge/Si$_{1-x}$Ge$_x$ on nanostructured silicon substrates. *Appl. Phys. Lett.* **2006**, *88*.

12. Carlin, J.A.; Ringel, S.A.; Fitzgerald, A.; Bulsara, M. High-lifetime GaAs on Si using GeSi buffers and its potential for space photovoltaics. *Sol. Energy Mater. Sol. Cells* **2001**, *66*, 621–630.

13. Wang, G.; Ogawa, T.; Soga, T.; Jimbo, T.; Umeno, M. A detailed study of H$_2$ plasma passivation effects on GaAs/Si solar cell Sol. *Energy Mater. Sol. Cells* **2001**, *66*, 599–605.

14. Shimizu, Y.; Okada, Y. Growth of high-quality GaAs/Si films for use in solar cell applications. *J. Cryst. Growth* **2004**, *265*, 99–106.

15. Wang, T.; Liu, H.; Lee, A.; Pozzi, F.; Seeds, A. 1.3-μm InAs/GaAs quantum-dot lasers monolithically grown on Si substrates. *Opt. Express* **2011**, *19*, 11381–11386.

16. Liu, H.; Wang, T.; Jiang, Q.; Hogg, R.; Tutu, F.; Pozzi, F.; Seeds, A. Long-wavelength InAs/GaAs quantum-dot laser diode monolithically grown on Ge substrate. *Nat. Photonics* **2011**, *5*, 416–419.

17. Lee, C.H.; Wang, J.; Kayatsha, V.K.; Huang, J.Y.; Yap, Y.K. Effective growth of boron nitride nanotubes by thermal chemical vapor deposition. *Nanotechnology* **2008**, *19*.

18. Bordel, D.; Guimard, D.; Rajesh, M.; Nishioka, M.; Augendre, E.; Clavelier, L.; Arakawa, Y. Growth of InAs/GaAs quantum dots on germanium-on-insulator-on-silicon (GeOI) substrate with high optical quality at room temperature in the 1.3 μm band. *Appl. Phys. Lett.* **2010**, *96*.

19. Liang, Y.Y.; Yoon, S.F.; Ngo, C.Y.; Loke, W.K.; Fitzgerald, E.A. Characteristics of InAs/InGaAs/GaAs QDs on GeOI substrates with single-peak 1.3 μm room-temperature emission. *J. Phys. D Appl. Phys.* **2012**, *45*.

20. Sandall, I.; Ng, J.S.; David, J.P.; Tan, C.H.; Wang, T.; Liu, H. 1300 nm wavelength InAs quantum dot photodetector grown on silicon. *Opt. Express.* **2012**, *20*, 10446–10452.

21. Tanabe, K.; Watanabe, K.; Arakawa, Y. Flexible thin-film InAs/GaAs quantum dot solar cells. *Appl. Phys. Lett.* **2012**, *100*.

22. Laghumavarapu, R.B.; El-Emawy, M.; Nuntawong, N.; Moscho, A.; Lester, L.F.; Huffakerb, D.L. Improved device performance of InAs/GaAs quantum dot solar cells with GaP strain compensation layers. *Appl. Phys. Lett.* **2007**, *91*.

23. Hubbard, S.M.; Cress, C.D.; Bailey, C.G.; Bailey, S.G.; Wilt, D.M.; Raffaelle, R.P. Effect of strain compensation on quantum dot enhanced GaAs solar cells. *Appl. Phys. Lett.* **2008**, *92*.

24. Popescu, V.; Bester, G.; Hanna, M.C.; Norman, A.G.; Zunger, A. Theoretical and experimental examination of the intermediate-band concept for strain-balanced (In,Ga)As/Ga(As,P) quantum dot solar cells. *Phys. Rev. B* **2008**, *78*.

25. Ilahi, B.; Sfaxi, L.; Maaref, H. Optical investigation of InGaAs-capped InAs quantum dots: Impact of the strain-driven phase separation and dependence upon post-growth thermal treatment. *J. Lumin.* **2007**, *127*, 741–746.

26. Nasr, O.; HadjAlouane, M.H.; Maaref, H.; Hassen, F.; Sfaxi, L.; Ilahi, B. Comprehensive investigation of optical and electronic properties of tunable InAs QDs optically active at O-band telecommunication window with (In)GaAs surrounding material. *J. Lumin.* **2014**, *148*, 243–248.

27. Azeza, B.; Sfaxi, L.; M'ghaieth, R.; Fouzri, A.; Maaref, H. Growth of n-GaAs layer on a rough surface of p-Si substrate by molecular beam epitaxy (MBE) for photovoltaic applications. *J. Cryst. Growth.* **2011**, *317*, 104–109.

28. Bollet, F.; Gillin, W.; Hopkinson, M.; Gwilliam, R. Concentration dependent interdiffusion in InGaAs/GaAs as evidenced by high resolution X-ray diffraction and photoluminescence spectroscopy. *J. Appl. Phys.* **2005**, *97*.

29. Wang, L.; Li, M.; Wang, W.; Tian, H.; Xing, Z.; Xiong, M.; Zhao, L. Srain accumulation in InAs/InGaAs quntum dots. *Appl. Phys. A* **2011**, *104*, 257–261.

30. Willis, S.M.; Dimmock, J.R.A.; Tutu, F.; Liu, H.Y.; Peinado, M.G.; Assender, H.E.; Watt, A.A.R.; Sellers, R.I. Defect mediated extraction in InAs/GaAs quantum dot solar cells. *Sol. Energy Mater. Sol. Cells* **2012**, *102*, 142–147.

31. Nozawa, T.; Arakawa, Y. Detailed balance limit of the efficiency of multilevel intermediate band solar cells. *Appl. Phys. Lett.* **2011**, *98*.

MDPI AG

Klybeckstrasse 64

4057 Basel, Switzerland

Tel. +41 61 683 77 34

Fax +41 61 302 89 18

http://www.mdpi.com/

Materials Editorial Office

E-mail: materials@mdpi.com

http://www.mdpi.com/journal/materials

www.ingramcontent.com/pod-product-compliance
Lightning Source LLC
Chambersburg PA
CBHW051921190326
41458CB00026B/6361